In Praise of *Dead Mars, Dying Earth*

"The difficult task of making real science writing read with the compulsive-ness of fiction is cannily achieved by Brandenburg and Paxson. Their straightforwardly compelling narrative manages to freight in a serious message about the destruction of the environment....This one is likely to join the ranks of major-selling scientific titles that cross over into popular appeal, and certainly carries a very salutary sting in the tail."

Publishing News

"This is ecological siren sounding on a cosmological scale, pitched at the non-scientist....[If] it is a writer's job to inquire about the world, then Brandenburg and Paxson have a considerable success on their hands: a timely and very frightening book, but one so engaging, it is more likely to inspire us to help save the planet than sink us—as more sober works too often do—into a cynical and useless despair."

SIMON INGS,
reviewer for Amazon.co.uk

"This is good! This is powerful! The science is informative, well written and nicely spiced and sliced with historical anecdotes and human dramas. I got started and could not put it down."

DR. HORACE CRATER, Professor of Physics,
The University of Tennessee Space Institute

"From a series of stories , a disturbing picture of a possible future emerges. Brandenburg and Paxson lead the reader to what seems to be an inescapable conclusion—that we may be leading our planet to a point of no return—and that something must be done, and done soon."

DR. MARK CARLOTTO, scientist,
Pacific Sierra Research

"I am, as far as I know, the only Director of a frontline international conservation advocacy organization, (among others, Greenpeace Foundation and Earthtrust), whose background is in the physical sciences and who actively champions the space program and what we can learn from it. So *Dead Mars, Dying Earth* really struck a powerful chord in me.

"With an action list that is in and of itself reason to buy the book, this compellingly-written and engrossing science drama paints a disquieting, plausible portrait of an Earth inching toward the brink of abrupt catastrophic planetary collapse, and challenges the reader to help transform human society past its current parasitic relationship with the planet that sustains it.

"I must say that as a trained geologist and scientific observer, I personally feel that the inclusion of the so-called Mars 'face' is entirely out of place here; but I feel the other Mars analysis offers powerful lessons and that the book stands strongly and lucidly on its many other merits. So, I hope that skeptical readers will, as I have, simply chalk the book's one questionable thread up to artistic license.

"The book's well-developed explanations of the existence of multiple 'stable states' for earthlike planets, such as those on burning Venus and frozen Mars, is an important and seldom-expressed concept, and one which we would be incredibly irresponsible to ignore. Perhaps most importantly, it is essential that we understand that planets can die, and this is underscored by recent NASA photos clearly showing the airless, dusty remains of a wet and once possibly living Mars.

"More generally, the broad range of arguments brought to bear on the book's other themes are impressive, and parallel my own thoughts, and I concur with all the book's recommendations. Also, although it is not a new opinion for me, I am scared spitless by what I see as the very real possibility of runaway greenhouse effects, and feel that we still simply know too little to discount its possible occurrence—in a case where hindsight would be of no use whatsoever.

"Read *Dead Mars, Dying Earth*. It's a transformational journey of science and passion you won't want to miss, and it just might forever change the way you view your world."

DON WHITE, Founder and President, Earthtrust,
Founder and Board member, Greenpeace Foundation,
also, Flipper Foundation, Project Delphis Research Lab,
Species Survival Network and the 2011 Exploration Society

"Fifty-five million years ago an abrupt, massive release of carbon from ocean reservoirs resulted in the warming of the oceans by 5 to 7° Centigrade. Extinctions occurred; and the Earth's ecosystems were forever altered. Could it happen again? In *Dead Mars, Dying Earth,* Brandenburg and Paxson propose a similar scenario. Are we now on the verge of a massive release of carbon? This compelling book 'digs' deep into the heart of carbon dioxide and other environmental concerns to offer both an inspired wake-up-call and an alarming warning."

JAMES ERJAVEC, M.S., Environmental Geologist & Geochemist,
Parsons Infrastructure & Technology Inc.

"Thank you for giving me the opportunity to read *Dead Mars Dying Earth.* I actually read the entire book in one sitting

"Not since the *Limits to Growth,* has a book so effectively made the case documenting the negative impact of man on the environment, while raising the specter of environmental catastrophe and yet holding out realistic solutions and hope. *Dead Mars, Dying Earth* goes one step further with a writing style that is less a documentary and more drama, holding the reader's interest and effectively making the case that our Earth is exceptionally rare and exceptionally fragile.

...As we hear news reports documenting the thinning of the ice shelf in Greenland and watch the forests being consumed by greed, important books like *Dead Mars, Dying Earth* are essential to raise the level of consciousness of the only ones who can change the trend—us!"

DON MAYER, founder North Wind Power Company
and owner of Small Dog Electronics
(a major Internet computer sales firm)

"This is the best book I've ever read! A meaty book, full of wondrous insight!"

JOEL BOWKER, Biologist,
Rustoleum Corporation

A fabulously alive, and deeply human tale of scientific revelation and dawning conviction, which ultimately charts a clear, sane path through an incredibly dangerous time for the health and life of our planet."

COMMANDER MARK HUBER,
US Navy

"I highly recommend *Dead Mars, Dying Earth*. It makes a clear and final argument: planets are fragile. Mars was once enveloped by a protective atmosphere and nurtured by flowing waters and this environment was lost forever. Unless we act now to solve the problems of human pollution, our planet could suffer the same fate."

"This is an intellectual tour-de-force. The authors range engagingly far and wide through the forest of scientific literature and disciplines to ferret out for us how Earth and Mars share a parallel ecological and geological history. Fortunately, rather than leaving us swimming in misery at the news, Brandenburg and Paxson give us the concrete steps to take to create a future Garden Earth, which will ensure that Earth avoids becoming a broken clone of dreary, dry and very dead Mars."

DR. JAMES STRANGE, field archeologist
and Professor of Religious Studies,
member of the Editorial Board of *Biblical Archeology Review*

"Paxson and Brandenburg are masters of telling a story... they scare you, excite you, invite you to laugh, but most of all, to think."

ULYSSES DOSS, Humanities Professor,
University of Montana

DEAD MARS, DYING EARTH

Dr. John E. Brandenburg, Ph.D.
and Monica Rix Paxson

THE CROSSING PRESS
FREEDOM, CALIFORNIA

First published in Great Britain in 1999 by
Element Books Limited
Shaftesbury, Dorset SP7 8BP

Contributing Editor, Stephen Kelley Corrick
Cover design by Max Fairbrother
Design by Behram Kapadia

Library of Congress Cataloging in Publication
Brandenburg, John.
    Dead Mars, dying Earth / by John E. Brandenburg and Monica Rix Paxson.
        p. cm.
    Includes bibliographical references and index.
    ISBN 1-58091-066-1 (pbk.)
    1. Climatic changes. 2. Environmental proteciton. 3. Mars (Planet) I. Rix Paxon,
Monica, 1951- II. Title.

QC981.8.C5 B72   2000
551.6–dc21   99-054229

The authors and the pubishers are grateful to the following for permission to reproduce
copyright material. Lyrics to 'Born to Be Wild' (M. Bonfire) reproduced by kind
permission of Manitou Music/MCA Music Ltd © 1968. Extract from *Dust Bowl Diary* by
Ann Marie reprinted by permission of the University of Nebraska Press; copyright © 1984
by the University of Nebraska Press. Text on 'How to Use a Peak Flow Meter' reprinted
with permission of American Lung Association; © 1999 American Lung Association.
Extract from page 49 of *Making of a Conservative Environmentalist* by Gordon Durnil
(Indiana University Press, 1995) reprinted by permission of Indiana University Press,
Bloomington & Indianapolis. Every effort has been made to contact all copyright holders
but if any have been inadvertently overlooked the authors and publishers will be
pleased to make the necessary acknowledgement at the first opportunity.

This book is dedicated to the memory of

Rachel Carson

and

Carl Sagan

Contents

Plate Illustrations

Colour Plates

1. Topography of Mars revealed by Mars Orbiter Laser Altimeter measurements taken on April 15, 1999 from Mars Global Surveyor. This image shows the Tharsis volcanic province and Valles Marineris. Shading was obtained by "illuminating" the topography from the north-east. The blue areas reveal the (formerly water-covered) Northern Plains.

2. Ozone depletion over Antarctica (1998). Coloured satellite map of atmospheric ozone in the southern hemisphere between mid-August and early October 1998. An ozone "hole" is seen over Antarctica.

3. Comet Shoemaker-Levy 9/Jupiter collision. Infrared image showing the fireball created as Fragment K of Comet Shoemaker-Levy 9 hit Jupiter at 10:18 GMT on 18 July 1994. Earlier impact sites are seen to the right of the fireball as glowing spots.

4. Chicxulub crater. Artwork of the Chicxulub impact crater on the Yucatan Peninsula, Mexico, soon after its creation. This impact may have caused the extinction of the dinosaurs and 70 percent of all Earth's species 65 million years ago.

5. Mosaic of images taken by the Viking I and Viking II probes showing the globe of Mars.

6. Sojourner. Mosaic image of the robotic Sojourner vehicle aboard the Mars Pathfinder. Image taken by the Imager for Mars Pathfinder (IMP) on the day it landed on the surface of Mars on July 4, 1997.

7. This mosaic of Viking Orbiter photos shows the boundary scarp between Mars' ancient cratered highlands and the northern plains to the east of the Mangalla Valles. Ancient river channels run north through the highlands. Notice the difference between the topography of the north and south. The latest NASA information indicates that the less cratered northern regions may have been the site of a paleo-ocean on Mars.

8. Forest fires, Brazil. The Amazon, Roraima State. The savannah area north of Boa Vista.

9. Erosion, Brazil. When the forest is destroyed, the top soil rapidly deteriorates.

10. Surface of Mars. Mars Pathfinder mosaic image of the dusty surface of the planet Mars. The area around Pathfinder features rocks such as the ones named "Flat Top" (at upper right) and "Wedge" (at upper left). Image taken by IMP on the third Martian day (Sol 3) it had spent on Mars since landing on July 4, 1997.

11. Cameroon. Aerial view of the Borombimbo crater lake close to Kumba in the South West Province. This is one of a line of crater lakes through the rainforest of Cameroon, one of which (Lake Nyos) released a deadly cloud of carbon dioxide killing 1,700 people in 1986.

12. Forest fires, Indonesia. Sumatra, Padang, Tanjung Bungus. Kerinci passenger boat stranded because of the thick haze.

13. Solar Power Station. Dish-shaped solar power reflectors at a solar power station at White Cliffs, Australia. Each dish focuses radiation from the sun onto a thermoelectric generator positioned in the focal point. A computer steers the dishes to ensure they face the sun throughout the day.

14. Fusion Research. Electrical discharges over the Particle Beam Fusion Accelerator during one of the accelerator's "shots." The target is a gold cone, at the focus of which is placed a small pellet of deuterium and tritium. The beam causes the target to emit very intense X-rays which collapse and ignite the pellet, causing a nuclear fusion reaction.

15. Forest Fires, Indonesia. Sumatra, near Bukit Tigapuluh. Kubu tribesman surveying the burning jungle. This area of forest has been used by generations for hunting and gathering medicinal plants.

16. Temperate Forest, France. Trees are the ultimate representation of life on earth. Through the planting of trees and the preservation of them, Garden Earth will be reborn.

Black and White Plates

1. Face on Mars. Viking 1 Orbiter photograph of surface features at about 40° north latitude on the planet Mars. These formations resembling a human

head measure 1 mile (1.5 kilometres) long. The photograph was taken on July 25, 1976.

2.   This is the Mars Global Surveyor image of the "face" that was broadcast on television and printed in newspapers around the world on April 6, 1998. It has become affectionately known as the "catbox" image.

3.   After processing to improve contrast and correct for smearing and skewing (the distortions created by camera movement and the highly acute angle from which the image was taken), this orthorectified image shows the same feature.

4.   According to James Erjavec and Dr. John Brandenburg, the line of demarcation between the knobbly terrain and the much smoother pediment surface (shown here in the Cydonia region of Mars) may indicate the shoreline of an ancient ocean or other large body of water.

5.   Egypt, Giza. View of the great Sphinx and pyramids. Could the face on Mars be a similar construction or is it simply an eroded mesa-like land formation?

6.   Joseph Priestley (1733–1804), the English chemist. He is best remembered for his discovery and examination of a number of new gases including hydrogen chloride, nitrous oxide, ammonia, nitrogen and carbon dioxide. He is often given credit for the isolation of oxygen although he never recognized its significance.

7.   The German bacteriologist Robert Koch (1843–1910). Koch is considered, together with Pasteur, as the founder of modern medical bacteriology. He was awarded the 1905 Nobel Prize in medicine.

8.   Svante Arrhenius (1859–1927), Nobel Prize-winning Swedish chemist at work in his laboratory. His achievements include his elucidation of the effect of temperature on reaction rates (Arrhenius equation) and the discovery of the Greenhouse Effect.

9.   Florence Nightingale (1820–1910), medical reformer. She developed the use of statistics, charting, graphic displays and management procedures in the practice of medicine and demonstrated that using a rational approach in an arena that until then had been left unmanaged resulted in a significant reduction in the number of patient deaths. Her contribution to the evolution of the modern medical model will be useful in helping us to manage the complexities of our planetary environment today.

Picture credits

The authors and the publisher would like to thank the following for permission to use color illustrations for which they hold the copyright: the Photo Researchers, Inc. for 1, 2, 3, 4, 5, 6, 7, 10, 13 and 14; Still Pictures for 8, 9, 11, 12, 15 and 16; and to the following for permission to reproduce black and white illustrations: the Photo Researchers, Inc. for 1, 6, 7 and 8; NASA for 2, 3, 4; Robert Harding Picture Agency for 5; Hulton Getty for 9. The Mauna Lao Chart on p. 274 is reproduced by kind permission of Dr. Michael Pidwirney. Every effort has been made to contact all copyright holders but if any have been inadvertently overlooked the authors and publishers will be pleased to make the necessary arrangement at the first opportunity.

Acknowledgments

Both of us are deeply indebted to the many people who have assisted us personally and professionally with this book. We would like to thank Caryle Hirshberg and Tierney Fox for their help. We would also like to thank the following for various contributions: David Alexander, Shelley Amdur, John Baldock, Kyren Bogolub, Diana Botsford, Ed Bouchard, E. Brandenburg, J. J. Brandenburg, P. R. Brandenburg, George Caledonia, Mark Carlotto, Alexandre Chahad, Michael Cipolla, Bruno Coppi, Barbara Corrick, Ernie Corrick, J. Patrick Corrick, Michael Corson, Horace Crater, Clarissa Cridland, Vince DiPietro, Ulysses Doss, Katalin Fekete, Sam Foertmeyer, Sonja Foxe, Ulli Genzler, Jeff German, Eva Ines Geweyer, Brenda Gill, Michael Goerden, Robert and Diane Guerra, Sarah Hadley, Thom Hartmann, Nancy Hill, John Kline, Brig Klyce, Amy Knapp, Jean-Claude Koven, Susan Lascelles, Laura Lee, Michael Mann, Gary McCabe, Julia McCutchen, Harry Moore, Bob and Barbara Munro, Peter Newell, James Redfield, Christopher Rollins, Francine J. Sanders, Martha Shaifer-Hartel, Robin Sheerer, Kirk Shelley, William Siri, Michael Smoler, the membership of the SPSR, Paul Taylor, Christy Tinnin, Dov Trietch, Johannes von Buttlar, and Sabine Weiss.

We would particularly like to thank James Erjavec, Anne Marie Tobias, and David Webb for providing invaluable support with our technical analyses.

Finally, our thanks go to Stephen Kelley Corrick, for his extensive contributions to the writings on trees, solar energy and the carbon cycle— and beyond that, for all he brought to the book as agent, editor, researcher, and friend.

JOHN E. BRANDENBURG
MONICA RIX PAXSON

Foreword

A new millennium dawns upon a humanity which, in a single generation, has seen both its technology and its knowledge of the cosmos expand like a newborn universe. Unfortunately, the new millennium also dawns upon a gathering storm.

Battle lines are forming now, battle lines for a conflict that could rend and will certainly define the first decades of the new millennium. Although this may appear to be mainly a scientific conflict, it is not confined to this arena. Instead, the conflict is primarily intellectual and perceptual—but its issues will cut to the heart of our civilization. Unlike the debate over continental drift—in which a geology professor's tenure might ride on how he perceived the relative shapes of Africa and South America—this debate will affect the energy budget of the planet and the economic lifeblood of nations. The front lines now forming are vast, stretching across sciences as diverse as marine biology and plasma physics, and through locales as far scattered as Amazonis on Mars and Amazonia on Earth. The conflict is intense, straining the veneer of civility in science to its breaking point and beyond. At issue is whether the Earth can die, and whether we ourselves are pushing the Earth's systems toward that end. Now it is not the ancient outlines of continents that cause debate, it is the looming outlines of inconceivable disaster: the Earth now marginally supports six billion human beings, but if its climate changes to any great degree, will that margin disappear?

The Earth can no longer be seen as the font of robust and limitless resources we once thought it was. We now know it is fragile and finite. To see our blue sphere from space is to recognize it as a fragile oasis in a harsh and indifferent cosmos. From space you learn that the oceans aren't boundless and bottomless. They're a thin slick of blue on a ball of rock that orbits the sun. Even the sands of the sea shore are numbered now. Before our eyes, the infinite becomes finite. Conversely, we are now in an age where environmental problems are no longer confined to toxic landfills or polluted streams, but instead involve resources and systems that are planetary—and when interests so vast are affected, conflict is inevitable.

At this writing, the sage and august board of the American Geophysical Union (AGU) has moved its 35,000 members to the battle line after a year's study of the scientific research. Representing one of the largest groups of scientists from the space and geophysical sciences, the Council of the AGU has now voted 26 to 0 to endorse a major statement, some of which is quoted here, and has pronounced the present course of human conduct unjustifiable.

> Atmospheric concentrations of carbon dioxide and other greenhouse gases have substantially increased as a consequence of fossil fuel combustion and other human activities . . . AGU believes that the present level of scientific uncertainty does not justify inaction in the mitigation of human-induced climate change and/or the adaptation to it.[1]

The AGU has thus thrown its enormous weight behind the scientific cause for the immediate environmental remediation of Earth. However, in a ritual that will soon become familiar, this grave and conservatively worded statement from the AGU was not greeted with thoughtful silence or even expressions of concern. Instead, it was the occasion for immediate cries of outrage and bitter denunciations from scores of ultraconservative, industry-funded organizations and political "think-tanks." These latter organizations are full of vague assurances and empty minimization of the risks of climate change to humanity. However, they implicitly raise the other valid question of this debate: how will we change the course we are on without terrible economic suffering? To confront climate change is to confront the energy-rich civilization we have built, and with it the whole Gordian knot of the human condition. The struggle is to find a path that avoids both environmental and economic disaster, a path that recognizes the worth of both a healthy biosphere and human progress. It is this balanced path that we hope to illuminate here.

So, we ask you to journey with us along the forming battle lines—from

Cydonia on Mars to Rondonia in Brazil—for only when you see the great issue of planetary life and death will you know the ultimate choices you must make and the side on which you will enter the battle for the life of the Earth. Global climate change, the depletion of the ozone layer, the Lyot impact basin on Mars, the burning core of Chernobyl and the heart of the Sun, are all way-points on this journey, the journey of the life and death of planets.

May your journey lead to life.

JOHN E. BRANDENBURG, PH.D.

This book is a bridge spanning a bottomless chasm. On one side of this abyss is our rational love of science and technology and our drive to create and explore. On the other side is our love of our children, our reverence for life and our utter dependence on Earth for sustenance. Beneath us is our past and an eternal fall from grace in which we may forever dance with the dinosaurs in a state of extinction. Above us lives fresh possibility, new life, the grand expanse of space, and visions of other worlds.

I hope to take you to the middle of the bridge. From there, no matter where you go in life, something will be forever altered.

The view from the bridge has transformed me personally. My life will never be the same. As a relative newcomer to planetary and environmental science, I started the research for this book from scratch, with a beginner's mind and nothing to prove. I spent months in total immersion, reading literally hundreds of scientific books, magazines, studies, reports, journals and abstracts, and visiting websites for more hours a day than makes any sense. But what I was discovering was so shocking, so amazing, so utterly unbelievable to me that everything else—other than my family, a few friends and this book—began to fade in importance.

In Chicago, where I live, there was a terrible tragedy in the 1950s in which dozens of children burned to death in a school fire, in part because the teachers hesitated when they smelled smoke, uncertain that there really was a fire, and therefore delayed ringing the alarm. It was a horrendous and fatal mistake in judgment.

The bottom line for me is this: anyone who thinks the science isn't there to support the notion that we are altering our planet's climate—to our own detriment—clearly hasn't read the mainstream scientific findings on the subject. Global warming is a dangerous reality. It represents a trend which, unchecked, will be as dangerous for all humans as that fire was for those children in Chicago.

I am pulling the alarm.

MONICA RIX PAXSON

Introduction

In a sense, we are all innocent. Like ants building on the side of a mountain, humans are simply an energetic life form aggressively colonizing the island-like landforms of Earth.

We share our earthly home with many other species, floating together atop a molten core on huge stony plates. We love and labor beneath a canopy of protective ether which both holds the warmth from our Sun and screens its deadly rays.

Just beyond our thin atmosphere lurks the frozen dark of infinite space. When our atmosphere retains too little of the Sun's warmth and light, we fight back by burning the tarry remains of ancient forests.

From the darkness of space we see that our planet is a watery sphere caught in the gravitational pull of a mid-intensity star. We share our star with a small group of planetary siblings. We proudly teach our children their names: Mercury, Venus, Mars, Jupiter, Saturn, Neptune, Uranus, and Pluto. But, do we really understand our solar system? Do we really sense the fragile balance of our own environment which is so delicately suspended between the extremes of Venus' blazing 800°C and Mars' frigid −100°C?

But, ignorant or not, we're adventurous children and we've boldly set out to discover the universe. We plan to scan and probe all the heavenly bodies we can mechanically place at the end of our far-reaching curiosity. We demand each planet tell us its story, reveal its hidden truth. Yet will we listen? Do we dare?

It is no time for timidity, because our nearest neighbor, Mars, with startling clarity, is revealing the story of its past—a wild and disturbing tale of life and death. More significantly, with the generous wisdom of an ancient warrior, Mars warns us of our future—a warning we should heed.

The story of Mars is on such a monumental scale that we are in danger of missing the whole point. Like ants unaware of the mountain, let alone the boulder rolling toward them, we can hardly imagine our place in the solar system, yet alone a threat from outside our range of perception.

Will we stand back and look? Will we hear? We should. We must. For what's at stake is our survival. Here before us is the opportunity to avert an almost inevitable planetary disaster—one that would make the Second World War, Hiroshima and Krakatoa seem trifling.

Surprisingly, it is our ability to explore space, and the lessons that we have learned there, that may allow us to save our planet. But knowledge is not enough. We must act, and act together. We are undoubtedly up to the task of healing our planet, but the limits of our present vision could prevent it. Yes, we have been innocent, but so were the dinosaurs. From here on we must be responsible or we, too, will perish.

The next step in human evolution calls for us to do something we have never done before—to act as a planetary community on behalf of our home world, protecting its environment and atmosphere. This is a complex task and we must accept its immense challenge even if we don't fully understand what's involved. We will learn.

This much is clear: there is no self-interest apart from what serves all of us. We will all share Earth's future, whatever its fate. The choice to act to save our planet is one we will need to make individually and collectively. To help you make that choice, please listen to the story Mars is telling us about Earth's future: the story of why Mars is dead and Earth is dying.

So begins the story of life and death on two planets, Mars and Earth. While this book may read like fiction, it is not. These are stories of science fact presented uniquely, unlike any other science book you may have read. The events this book describes have actually happened—or may well happen. Even the future events we discuss have also been put forward as serious, well-developed hypotheses in published scientific papers and journals. So, although we talk about dramatic events, both wondrous and terrible, this is not a movie script nor wild, quasi-scientific speculation.

Though unquestionably sobering, the contents of these pages also speak of our long-term future on this planet, and offer real solutions for our difficult problems and fresh possibilities for humankind. In this book we'll discover how Mars, the bringer of war, helped end the Cold War here on

Earth; how Mars shared a warm, wet past with our own planet; how it lost the ability to support life, and how understanding the death of one planet may help save the life of our own.

CHAPTER 1

Between Fire and Ice

Some say the world will end in fire,
Some say in ice.

FIRE AND ICE
ROBERT FROST

The acronym "OID" in big letters dominated the title sheet I tacked to the bulletin board as I set up my paper at the American Geophysical Union (AGU), a major international scientific conference. I was here to deliver a paper, in this case a poster paper—a paper delivered in written form on a bulletin board—in a room full of poster papers. My name is John Brandenburg. I've presented quite a few scientific papers at poster sessions with all sorts of results—from passionate shouting matches to indifference—but the reaction to this paper was to be unlike any other I had experienced.

OID, for Oxygen Inventory Depletion,[1] is a disquieting discovery I made with my co-author Monica Rix Paxson and our associate Stephen Corrick in the process of writing this book. It predicts, based on increased carbon dioxide in the atmosphere, that Earth's oxygen levels are dropping. Our findings were presented before the AGU on December 10, 1998, and were the first published work to openly make this prediction, and I could tell right away that we'd struck a nerve in the scientific community.

The acronym on the title page drew immediate attention from the passers-by and, just as I tacked up the second page, one fellow (whose nametag said he was from "NASA HQ") stopped and stared intently.

Frankly, the paper was short and to the point, no more than what was required: some equations, a few statements with bullets, a risk/benefit matrix, one graphic and a couple of footnotes. As he read, his eyes got large. I finished hanging the final page and turned to explain the details of our analysis. "That's very alarming!" he blurted abruptly and then dashed away, almost as if he'd seen a ghost. I watched as he fled, trying to determine if he was rushing to tell someone about the paper or perhaps simply could not accept what he had just learned. I couldn't tell which, but felt suddenly chilled. I knew I was the bearer of bad news, but I hadn't expected a reaction as intense as the one I'd just witnessed.

Soon others replaced him, all staring intently: students from the world over, from India to Indiana, scientists from China and Europe. I would have found the level of interest gratifying except there was no easy banter, no friendly opposition. I would have explained our analysis but no one questioned our conclusions or argued with me. Instead, they nodded sadly as if acknowledging the truth of OID's stark prediction, a level of acquiescence I hadn't been expecting.

"OID," one poster visitor said aloud as if tasting the sound. "It sounds like Oi Vey," he joked without smiling.

"Oi Vey" is right, I thought, thinking how appropriate the all-purpose Yiddish expression of angst was in this case.

"It's obvious in retrospect," observed a student from the University of Ohio after carefully considering the paper's significance. "I wonder why no one has measured the oxygen levels?" he asked.

"I believe the observatory at Mauna Loa already has," I replied. "When we first made the discovery six months ago, I called them to ask for measurements. When I asked if there was a decrease in atmospheric oxygen, the response was, 'What do you think?' Beyond that they suggested we should figure it out on our own. We already had. Who knows why, but I guess they were keeping the data to themselves. When they didn't publish their findings, we thought it was important to bring our findings here to tell the rest of the community about OID."

Soon thereafter a tall scientist from Denmark arrived at my poster, looked at the paper and smiled. "It's been measured," he said, nodding with satisfaction. "I saw the data presented at a closed-door meeting two months ago. They were pleased that the drop matched their predictions."[2] Considering that we were talking about a potentially devastating piece of news, his attitude might have seemed blithe. But that is the ironic satisfaction one sometimes gets as a scientist—like the empty satisfaction the designer of the *Titanic* must have felt while sitting in the ship's ballroom when it sank in exactly two hours, as he had predicted: assured of the confirmation of his finding in spite of the bad news that he was still on board.

"I consider our model conservative," I suggested to the Danish scientist. "It says the level has declined by 50 to 70 parts per million, not enough to harm humans yet, but still a dangerous trend. What do their measurements say?" I asked, suspecting the loss of atmospheric oxygen might be worse than we'd calculated it to be.

The man from Denmark just looked at me and smiled with tolerance, nodded and left without another word. At least the Danish scientist's reaction had been pleasant. Six months earlier, at the Spring AGU meeting, in a similar poster session, I had been one of a group of scientists who had presented our analysis of the Mars Global Surveyor images. As a senior scientist on the MGS mission approached our poster, he flew into a rage at our analysis, and threatened a member of our team with his fists. Mars images, it seemed, were very important to some government officials. Now, again, very unexpectedly, it was looking like analyzing Earth's oxygen level might be that way, too.

At one point, a woman who'd stopped at our poster suggested that I come with her. "I'll show you why this is happening," she offered. Curious, I excused myself and followed her to a much grander color poster several spaces down, where her own scientific findings were being presented. In the center of the display was a large satellite image—a mosaic-like map of the Amazonian basin. "Here, this crescent," she swept her hand across the bottom of Brazil in a vast arc from the coast to the Andes, "used to be rainforest but it's all gone now. Thirteen percent of the Amazon is deforested now."

"I've heard that there are places in Amazonia that used to be forests but now are nothing but cracked red earth for as far as the eye can see," I ventured, hoping my description was an exaggeration. "They say it looks like Mars."

"Yes," she replied matter-of-factly. "That's here in the south, in of the state of Rondonia. It's hosed." She pointed to a region outlined with the provincial boundary.

"Hosed?" I asked, unfamiliar with the vernacular.

"Wasted," she explained. "It's a wasteland now; a desert."

I stared at the photographic evidence in stunned silence, suddenly getting a taste of how the readers at my own poster session must have been feeling.

"That's near where we work, actually," she continued. "It's very dangerous there. The land is devastated and the farmers and ranchers are desperate. They've been known to kill people from the outside who come asking questions about deforestation. A few months ago we heard that a group of reporters were coming, so our whole team scattered until they were gone."

I looked at her with admiration. She was very brave to research where she does, a personal characteristic I admire. "How did you discover OID?" she asked.

"On Mars there's a region called the Amazonian Plain," I replied, "just like here on Earth. It's a vast desert, too. Ironic, isn't it? I know it seems strange, but we discovered it there—at least that's part of the story. Believe me, it's been quite a journey ..."

Get your motor running
Head out on the highway
Looking for adventure
In whatever comes our way

Lyrics to "Born To Be Wild"
by M. Bonfire

On a sunlit afternoon in August 1972, Michael Rose and his girlfriend Sherry were cruising a winding mountain road in his hand-painted VW microbus, enjoying Idaho's rolling mountain vistas spiked with blue-green firs and pines. They were relaxing, laughing and relishing the perfect freedom of a day with no certain destination. Like so many of their generation, Michael and Sherry may have been looking for adventure, but they never could have dreamed of what actually came their way. As they approached the crest of a ridge, a second sun appeared on the southern horizon. In a heartbeat, Michael slammed on the brakes, threw the car into park, and jumped out the door. Standing dazed beside his embellished vehicle, he was witness to the unbelievable—a giant, pulsating meteor was headed straight at them.[3]

"It was really *big*," Michael said, his frustration at trying to explain what he saw evident in his voice. "I don't know how big, but maybe something like a football field, and it seemed real close. *Way* too close." Michael was right about that. It was far too close. Although the hurling hunk of cosmic matter was later estimated to have passed within 36 miles of the Earth's surface, on the cosmic scale that is but a mere fraction of an inch.[4] Had it been a bullet pointed at your head, you'd have had a permanent parting in your hair.

"At first I thought it was some kind of spaceship or something like that. But that didn't make any sense because it was flaming," Michael reasoned. "It glided overhead from south to north, like some sort of huge jetliner coming in for a landing—only it was a giant ball of fire. And the colors were amazing—red, orange, yellow, green, blue—rainbow colors that kept changing. And that's not all. It had a flaming tail that left a thick, pinkish-orange vapor trail behind it."

In eastern Montana, Mrs. Jack Davis was timing youth club events at the Park County Fair when she shouted over the public address system to her astonished audience, "Look over there, kids! It's a fireball. You might never see one of those again in your whole lifetime."[5] Dozens of children focused on the passing meteor as it slipped silently across the sky. Although the fairgoers watched in unabashed amazement, the world's astronomers were less taken by surprise. This meteor, though far larger than most, was actually part of an anticipated annual astronomical event—the Perseid meteor showers—the yearly return of a brilliant mass of meteors, all remnants of a comet that disintegrated in 1862.

Whether it was anticipated or not, Dempsey Johnston, a forester who was at the Spotted Bear Ranger Station near Kalispell, Montana, took the arrival of this particular Perseid meteor in his neck of the woods very personally. "I was about ready to dig a hole. I didn't know what was happening. It looked like it was coming right at me."[6] It was coming at him all right. Fortunately, it flew right over and didn't stick around. Only a long trail of smoke remained, lingering for more than an hour.

Captain Bob Bagshaw, piloting a flight from Missoula, Montana, to Salt Lake City, Utah, for Frontier Airlines, was amazed when the meteor appeared in his field of vision. Captain Bagshaw had a ringside seat. "I've never seen anything so impressive in my life. It was a big fireball that was flaking off particles and leaving behind a reddish-orange trail. It started on the horizon in the south, went right across my windshield and out of sight on the horizon in the north."

Fortunately, the Federal Aviation Administration employee was inaccurate in his estimate when he told area newspapers that the boxcar-sized object was at an elevation of about 80,000 feet when it crossed just east of Missoula.[7] It would have caused the equivalent of an atomic explosion had it impacted the Earth, and it came very close.[8] Had the meteor entered the atmosphere either lower, slower, or at a slightly more acute angle, it would have been captured by the gravitational pull of the Earth, resulting in an explosion up to five times greater than the Hiroshima atomic bomb. While some might call this meteor's pass through Earth's atmosphere a near miss, a near hit would be more accurate.

However, it was the very real possibility of an atomic explosion of another kind, the kind produced by a nuclear attack, that alarmed the command on duty deep within the steel and concrete bunkers of the North American Air Defense (NORAD) buried in Cheyenne Mountain, Colorado. In a state of perpetual high alert during the Cold War, NORAD scrutinized anything that appeared out of the ordinary in the skies (in other words, anything that wasn't one of the various orbiting satellites they normally tracked by radar.)[9]

What appeared on the radar screen that August 11 must have been particularly alarming. Clinging evasively to the curvature of the Earth, something was screaming in from the south—the direction considered the most likely for a surprise nuclear attack. Was the United States under attack? Clearly. But by what? Everyone present was galvanized by the potential threat. Any decision to defend or retaliate had to be made immediately. A misreading of the critical evidence—either way—would be disastrous.

Fortunately, NORAD correctly identified the flaming object flashing across the sky and did exactly the right thing: nothing. So, the Perseid meteor moved swiftly on to escape the Earth's atmosphere above Canada, 1,500 miles from where it first entered. During the course of its brief visit it caused no damage apart from fraying the nerves of those who witnessed its presence and evoking the horror of the atomic explosions that might have been set off had it been a bit closer to the surface of the Earth or misidentified as an nuclear attack.

Having caused no war, no explosion, no gaping hole in the surface of the planet, this near-death meteor vanished from sight. It slipped back into the vast darkness beyond our atmosphere, swallowed once again by the endless, penetrating cold of space, its flaming rainbow aura forever extinguished.

But this meteor left behind more than a 1,500-mile streak of notoriety. It left a grim reminder: that our Earth is like a duck in a very busy shooting gallery. Whether we are aware of it or not, we are bombarded daily by cosmic matter of various sizes, shapes, and materials. Ordinarily, the buckshot that fires in our direction is harmless. It falls to the surface of the Earth occasionally, but more typically burns up in our atmosphere, or even bounces off like a stone skipping on water. However, every now and then, a cataclysmic strike forever alters life on this planet, sometimes by nearly wiping out life altogether.

While we have just begun to search space for "incoming" threats, and don't yet have a method in place to prevent their impact, we do have a first line of defense: our atmosphere. Our atmosphere is a thin coating of gases that adheres to the surface of our spherical home. From our Earth-bound perspective, our bright sky seems to reach endlessly upward, but is, in fact, no thicker than an apple's skin is to an apple. Still, a thin skin is better than no skin. We have only to observe our Moon or the planet Mars to see why. Neither has an atmosphere, or very much of one, and both show evidence of cosmic bombardment. Thousands of overlapped, interpenetrating craters pockmark their surfaces.

It's not that Earth hasn't been hit by meteors. It has. But it has nowhere near the tin-can-left-too-long-in-a-shooting-gallery look of Mars or the Moon. The oceans of Earth and the processes of weathering account for some of the missing evidence of cratering, but the protection our atmosphere provides is the primary reason our planet has endured fewer strikes. Relative to the vacuum of space, the gases of our atmosphere are extremely dense. When asteroids, comets, or meteors first contact Earth's outer atmosphere— the stratosphere—at a fairly flat trajectory, they are likely to ricochet off in another direction. If they aren't deflected, they are superheated upon entry by friction with atmospheric gases and typically burn up (more or less completely). In the vast majority of cases, we are spared serious impacts and the craters they leave.

The blanket of the atmosphere provides our Earth with other forms of protection as well. For example, the insulating property of the atmosphere holds in the warmth of the Sun and keeps the aching coldness of space at bay. It also protects us from being fatally irradiated by the rays of the Sun. The refractive properties of the gases in the atmosphere cause the light of the Sun to diffuse and bounce around, creating the great blue arch of sky over our heads. You also might argue that the atmosphere protects us, at least during daylight hours, from the visual impact of our true circumstances in the universe—from the fact that we are on a tiny planet orbiting a mid-intensity, G-class star that is the sole source of heat and light energy for a world falling endlessly through a vast void of frigid darkness.

Because we can't always physically see our cosmic circumstances directly, it helps to use imagination in order to assemble and assess the facts, and then to project them into the future to begin to understand their significance. For example, imagine this: without our atmosphere, we would be like Mars. Pockmarked and cold. Very, very cold. And dead.

The ability to imagine can be essential to survival. It isn't adequate to have only facts. Sometimes we have to be able to project circumstances into the future in order to see what is significant and what isn't. We may also want to consider the accuracy of the facts we are already very confident of, lest an essential element that previously escaped our attention be thrust through the threshold of our awareness as suddenly as a meteor.

"The trouble with him was that he was without imagination. He was quick and alert in the things of life, but only the things and not the significances." These lines from Jack London's story "To Build a Fire" describe a man who freezes to death unnecessarily in the subzero Yukon.[10] The tragic "chechaqou" (tenderfoot, a newcomer, unused to hardships) dies from a fatal lack of imagination. Overly confident, the man strikes out alone with no backup and no source of advice—except for a dog. If he had been traveling with an seasoned Inuit, he probably would have lived. Most likely, the Inuit would have looked at the sky and then at the tenderfoot's attire, imagined the consequences, and said, "This trip today is not a good idea. Let's make camp here and have some tea instead." The Inuit know that life is precarious and that a sunny, blue sky can kill just as surely as a blizzard. In fact, a blue sky can kill more swiftly because it holds less of the Sun's heat, and the planet's surface becomes colder—like outer space.

So, this is our perpetual dilemma: because the Sun is hotter than a welding arc and outer space is cryogenically cold, our lives are perched in a razor-thin zone between these extremes—with the slightest tip in either direction enough to dramatically alter our reality.

Humans require conditions for life that are exceedingly rare and frail in our cosmos. Most of the matter in the universe exists either in a plasma state (so hot that even atoms dissolve into a gas of electrons and ions—11,000 °C above absolute zero and higher) or as an incredibly rarefied vapor at the temperature of liquid helium (three °C above absolute zero). Poised between searing heat and biting cold, life thrives in a narrow zone of thermal balance. While we might find other species some day that can regularly sustain life outside the narrow range of Earth's temperatures, humans are not able to do this.

We are like an Inuit, who by wisdom and imagination crafts an igloo from a featureless field of snow and dwells in it. The igloo is kept warm—but not too warm—by a fire in its center and by the heat of his own and others' skin. Too much warmth and the igloo melts, leaving them without shelter. But, on the other hand, if the fire fails, the igloo becomes an uninhabitable cave of ice. As any Inuit who practices the art of igloos will tell you, to dwell in an igloo in winter requires skill and realism: a sense of balance. You must have the ability to balance heat and cold and not be too demanding of warmth. No roaring fires on the hearth here. Your shelter is made of ice crystals, so the walls must stay below freezing while your small fire burns in the center, and your body must be at some temperature in between. It is thus recognized among the Inuit that your shelter has needs as well as you do. If you abuse its hospitality, it will collapse and you will freeze. You must make careful choices and have both imagination and a sense of humor about your situation; you also need to respect that you live balanced on a knife's edge where hot and cold cut both ways and human flesh is very tender. Life on Earth is like life in an igloo, with little margin for error.

Balance is the appropriate term because everything in the cosmos is part of a dynamic and ongoing process involving interactive change, like the skillful way that a tightrope walker manages her body to stay aloft a slender cable. Thermal balance in the cosmos is determined primarily by radiation. Objects in space, such as planets or asteroids, absorb energy from the light radiation of stars. In the case of most stars, the major part of this radiant energy is in the visible wavelengths. That's why we say stars "shine." When other objects absorb the light from stars, their temperature rises and they too radiate energy. However, since planets are colder than the stars, planets radiate in the infrared range—primarily as heat—instead of in the visible range. When they have received enough warmth from the light of a star, objects in space radiate as much energy per unit of time in the infrared as they have received in the visible light range: thus thermal balance is achieved. When the rates of energy absorption and radiation reach a balance, the temperature of the object stops increasing and becomes constant. If this balance results in surface temperatures above freezing and below the boiling point of water, then life as we commonly know it can exist. While we will find a few exceptions to this rule—life forms growing in boiling water issuing from vents at the ocean floor or hidden deep inside the ancient polar ice of Antarctica—human life is not exceptional. We must live in the middle, away from the extremes of fire and ice.

When we look to the skies and spy on our nearest neighbors, we quickly appreciate the kernel of wisdom in the old cosmic real estate adage: location, location, location. Venus, wrapped in its heavy atmosphere of carbon dioxide, is closer to the Sun than the Earth is, so it is too hot for habitation. Only 50 minutes after soft-landing on the surface of Venus in 1972, the Soviet craft Venera 8 probably burned to a cinder in temperatures that its own instruments measured at 760 °C before quitting. As if its heat wasn't extreme enough, the atmospheric pressure on the surface of Venus, as recorded by Venera 8, is 90 times that of Earth's at sea level.

At the other extreme, Mars, with its thin but active atmosphere of carbon dioxide, is farther from the Sun's unrelenting heat than Earth is. Its atmospheric pressure is less than 1 percent of Earth's, so although Mars is better than Venus as a place to evolve and sustain life, it is too cold for most of the life forms we are familiar with. Like baby bear's porridge, only Earth is just right for us. While these planets bracket our Earth in terms of temperature, one hotter and the other colder, it is interesting to note that the same gas—carbon dioxide—makes up the atmospheres of both Mars and Venus. High carbon dioxide levels are an atmospheric signature of planetary disasters that have run their course. ›

Distance from the Sun is only one factor determining the thermal balance at a planet's surface. Temperature is also determined by the mixture of gases in the atmosphere and the materials of the planet's surface. Some of these materials absorb heat and others radiate it. Absorbers of heat absorb the visible range of light energy, whereas radiators of heat radiate in the invisible (to us) infrared range. Visible and invisible ranges of the light spectrum are so different that the processes of absorption and radiation are often dominated by different substances. Water is the most extreme example of this. It is transparently clear within the visible spectrum, but registers black as ink in the infrared and ultraviolet ranges, which are invisible. Earth's soil, which includes a great deal of organic matter, absorbs in the visible range and radiates heat, but radiation of heat from the Earth is primarily dominated by atmospheric gases—water vapor in particular. So it could be said that the thermal balance of the Earth involves a fine interplay of diverse elements interacting in symbiosis. It is a diversity that acts cooperatively.

Earth's biosphere (and possibly other biospheres we will find) supports myriad communities of organisms that achieve symbiotic balance. On Earth, the most basic of these is the plant/animal symbiosis—the most essential form of Earth's biodiversity. Plants take in carbon dioxide and water and form sugar and oxygen. Animals consume both sugar and oxygen and make carbon dioxide and water. This system, when operating as it typically does on Earth, creates an oxygen-rich atmosphere and an ozone layer, which in turn allow animal life to exist on land as well as in water. The system equally provides adequate carbon dioxide for plants to grow both on land and in water. Thus, the cycle is self-replenishing and a balance of gases that support all forms of life is maintained. As long as this system maintains its balance, life on Earth can flourish. If the balance of the cycle is disturbed, life is disturbed as well.

For those of us whose secondary education started in the mid-1970s or later, the concepts of ecological systems and balance in the environment were probably first introduced in biology or science classes and the theme of environmental balance will thus be familiar. But before the '70s, these concepts were not widely taught or understood by anyone other than a few experts. People in their 20s and 30s today are from the first generation that has been educated in school about the complexities and struggles of our relationship with the environment; they are the first generation to be educated about living on the knife's edge as a day-to-day reality. (Of course, countless earlier generations learned about this when their crops failed or the hunt was unsuccessful.)

But knowledge does not always bring power. The near quarter-of-a-century onslaught of ecological woes has not given most young adults the gift of a fighting spirit ready to save the environment; rather, they have a deep-seated sense of cynicism or despair about the condition of the planet. They fear that we are doomed. Their attitude is similar to the grinding fatalism with which the previous generation faced the day-to-day possibility of nuclear war. The generation before that had to live with the harsh realities of world war and economic depression. It seems as if every generation is handed at least one life-threatening problem not of its own making to solve. Regretfully, today's young are no exception. Like Michael Rose and his girlfriend, whose leisurely drive almost got them hit by a meteor, we're all looking for adventure, but what comes our way is often not what we're hoping for.

What's scary is to see how immobilized those trained in environmental awareness have become. While there are also many unnecessarily resigned older adults, with so many young adults, teenagers, and children personally resigned to ecological Armageddon, it's important for those of us who didn't grow up with environmental horror stories to understand the overwhelming despair that comes when you truly realize the enormity of the problem. As a young man recently put it, "Ask me anything about the problem of disposable diapers. I know all about that. But what am I supposed to do about everything else? I think it's a lost cause here. We'll probably have to go to the Moon or to Mars."

So, for many of our children, the prospect of leaving the Earth for another heavenly body in the neighborhood seems like a viable alternative to staying here and fighting it out on behalf of life on Earth. Fortunately, the assessment that our environmental situation is hopeless is not only premature, but it also ignores our capacity to courageously take on challenges. Humanity has repeatedly proven that we are fully capable of rebuilding and recovering from both devastating wars and environmental disasters. Is our situation hopeless? Probably not, but we'll never know unless we exercise the full force of our talents to make things better. Besides, we'll have to get this ecological balancing act right *before* we colonize space or our attempts will be very short and very deadly. The unforgiving extremes of space require complete mastery of our environment in order to survive. If we can master a healthy relationship with this planet, we will undoubtedly be able to move on to others.

Indeed, it takes incredible courage to face what isn't working on this planet. It is an individual challenge—to personally make a difference by responding to our planet's needs. But when we each courageously decide to take on the challenges, we find that our numbers need not be great to produce great change. As Margaret Mead said once, "Never doubt that a small group of thoughtful, committed citizens can change the world. Indeed, it's the only thing that ever has." Later in this book we will certainly talk more about changing the world. Meanwhile, you should know that there is a good deal of historical evidence that says that humans are capable of positively transforming both the Earth and themselves—even when that transformation at first appears to be impossible. But before we get to that, there is more for you to hear and ponder about. Like so much that is beautiful and worthwhile, the stories of Earth and Mars are complex, and deeply interwoven with human concerns. More than scientific facts are required by those of us who would survive—lest the power of our own dynamic thirst for life be squandered in the arid landscape of textbooks.

We will make sure that you find more than dust to drink here, because one cannot explore the inside of a star, ponder the future of human evolution or even recognize the significance of a scientific truth without the context provided by a story. Perhaps the best place to begin is with the story of a young man who crossed the threshold into adulthood in a very troubled time not of his making, the early 1970s.

I want to tell you the story of the first time I saw "the world." I was a student in college when my grandmother Edith died. When she departed this life, she left me a very generous and unusual legacy. She willed each member of my family money, along with the provision that the money be used to travel. Her gift was the extraordinary opportunity to see something of the world outside of Medford, Oregon, the small city founded by the lumber industry where I'd grown up. So, in the years following her death, the members of the Brandenburg family became world travelers for the first time. Each of us set off in a different direction according to our compelling interests.

My parents, John T. and Muriel Brandenburg, who were devout Christians, visited Israel and Greece, and were thrilled to see the places they had studied in the Bible all of their lives. My brother, who was interested in culture, went to Europe, as did many young Americans during the early 1970s. My decision ran contrary to the impulses of most of my generation, however. Motivated in part by a deep curiosity that has often led me off of the beaten path, as a 20-year-old and a second-year student at Southern Oregon State College, I resisted the many allures of Europe's warm friendship and decided to spend my legacy on traveling deep within the enemy's embrace. Part wary soldier-to-be, part hopeful pilgrim, I decided to go to Russia.

During the height of the Cold War, I headed off from Alaska on a warm summer's day in an Alaska Airlines jet, destined for Siberia. In a sense, it was a self-assigned mission to study the "enemy"—to get to know this powerful opponent of my country on a first-hand basis and to gain both knowledge and wisdom—because I fully expected to be involved in the defense of my country against this foe someday.

The expectation that I would be called to protect my country was partly my family's legacy. My granduncle fought in the First World War and my father fought in the Second, as did my three uncles on my mother's side. That I would serve my own tour of duty seemed inevitable—something that I would naturally be willing to do as part of what was required of a good citizen. Besides, I was very much living in a time of war—the Vietnam War—and inside my wallet, in the pocket of my blue jeans, was a draft-card bearing a number that had been assigned to me by lottery: 55. I liked the lottery. It was simple and it was equitable.

At the time, healthy young men of my age were being drafted into the American Army, trained, and sent off to fight and, too frequently, to die in Southeast Asia. When I went to Russia, I fully expected that those from my hometown who had been assigned numbers up to 75 would be drafted. Although it wasn't an obvious or direct route, settling the world's major conflicts seemed like my best chance to survive—the best chance we all had. So Moscow seemed the most appropriate place to spend my summer vacation.

My first destination, Khabarovsk, near the fateful juncture of the Amur and Usurri rivers, on the Sino-Soviet frontier, was inside a vast, 6-million-square-mile region that spans much of the north of Russia. Siberia is so large, in fact, that it could easily contain the whole of western Europe. Its terrain alternates between mountains, high plateaus, and plains; and the climate ranges from frigid Arctic and treeless tundra to the vast, scrubby grasslands of the steppes. One of Siberia's most remarkable features is the primeval forest of larch and evergreens known as the northern boreal forest, or, as it is known in Russia, the taiga. The taiga is part of a worldwide system of forests that ring the northern hemisphere just below the Arctic. The boreal is the forest of fairy tales and childhood dreams—where wolves lurk in the shadows and Snow Queens reign, and dainty arctic foxes, dressed in their very own white winter furs, prance in the snowdrifts. This was also where, under a cold blue sky, clad in snow-white winter camouflage, with bayonets gleaming, the world's two largest armies, Russia's and China's, faced each other with murderous stares.

I learned my first lessons about Russia from the window of the plane. There, parked on the side of the landing strip in Siberia, was a herd of mammoth, monster tractors sporting bumpy, bulbous tires that were taller than I was. Even though it was mid-summer and the weather was warm, the huge plow blades attached to the front of some of these tractors told another story: the story of massive battles with snow and ice. Although I lived in a part of the world where winter was cold, I had never seen anything on the scale of this equipment. Siberia is a very cold place, I concluded. In fact, the forests of Siberia are frequently captured in a long, deep, winter freeze with temperatures dropping to –40 °C. I also saw through that window the legacy of other battles. Khabarovsk appeared to be a vast, drab-green military encampment.

Not 20 miles from Khabarovsk, in the spring and summer of 1969, Russian and Chinese militaries had clashed bloodily in three battles of successively greater scope and violence. My guides shuddered with dread as they recalled the conflict. The immediate focus of this savage fighting had been a wretched mudbank in the Amur river, submerged most of the year, called Damansky by the Russians, and Chen Pao by the Chinese. In the wake of those battles, and others that raged all along the Siberian frontier that summer, the Russians had rushed millions of troops and millions of tons of equipment to face their one-time Communist ally. I had arrived in Khabarovsk as it was being transformed into a fortress.

I learned another lesson about Russia from the window of a second plane. As we approached the city of Moscow from the east, the city could hardly be seen. It was covered by a dense, smoke-like haze. When I asked the two women acting as cross-country tour guides about this, I was told a fascinating account of how Moscow was surrounded by underground fields of peat, some of which had been slowly smoldering while continually thwarting efforts to put out their hidden fires. Evidently, this had been a persistent problem, and knowledge of the situation was undoubtedly widespread, since the smoke was clearly impacting air quality everywhere in the city.

Later, on a tour of Moscow, we were escorted by a different pair of guides who were experts on the city. The smoke was very bad that day, burning our eyes and choking our lungs. When someone from our party complained to our new guides about the lingering smoke, we got an altogether different response than the story about burning peat. "What smoke?" was the response. "There is no smoke." The stunning denial floated through a choking haze of the obvious.

Apart from the form of smoke-induced blindness that seemed to affect some citizens, the people of Russia, I discovered, were pretty much like the ones I knew back in Oregon—warm, intelligent humans who cared about their families and loved their country. They had the same kinds of concerns, too: about work, money, and housing. However, unlike the smiling faces I saw at home on the streets of Medford, the Russians were serious, almost dour, as they walked down their city streets.

The only exception I saw to this general gloom was in Leningrad (now St. Petersburg). An emaciated looking man was running as fast as he could toward me as I was out for a walk on the city's streets. His expression was one of unmitigated terror. As he ran frantically past, he glanced backward over his shoulder as if to see if he was being followed. The color was drained from his face and I could sense his deep panic. He ran past me, and I, suddenly, as if by reflex, concentrated on looking as if nothing had happened and walked on.

Soon after the running man evaporated into the distance, a muscular truck pulled in to the curb. A group of men, civilians wearing red armbands and wielding billyclubs, jumped off the back and began searching the surroundings. Although I could only speculate about what I was witnessing, in that moment I had a taste of the fear so much of the world experiences daily in the throes of totalitarianism, and I knew that I'd just encountered the threat I had come so far to study.

Some threats are overt and visible, whereas others are not: they are entirely invisible to normal perception. Some threats pursue noisily, waving billyclubs and guns. Others are transparent, silent and unarmed, yet may be equally deadly. Disguised as your friends, bearing gifts that seem to have no cost, some leave you swimming in a tasteless, odorless, invisible ocean of wastes. Others float hidden in plain sight, like the choking smoke in Moscow, robed in denial. We fear the enemy of totalitarianism, but another kind of threat, an environmental one, can also grab its victim in a deadly stranglehold. This enemy lies waiting everywhere on Earth, respecting no border, no civil authority, finding us even in the warm shelter of our homes, *every* home, even those that seem safely tucked away in the very heart of a wealthy democracy. Just ask Christy Ann Booker.

W hen Christy was two years old, an unusual noise began to come from her throat—a slight gasping noise. Christy's mother, who had been decorating the family home, located just outside of St. Louis, Missouri, stopped her work to listen with rising alarm to the sound of little Christy's labored breathing. The wheezing was the sound of bronchial tubes starting to close down; the vibration was her two-year-old lungs striving to pass air through the pinched tubes that connect them to the outside world. Christy was having trouble with life's most essential function: breathing.

Until then, Christy's life had been remarkable only in that it was so normal. She was born on June 8, 1970, after a routine delivery, weighing a healthy 7 pounds 8 ounces. Soon after birth, Christy was released from the hospital to join a world already in the full force of the latter part of the twentieth century. At the time of Christy's birth, there were 3,697,141,000 people living on Earth, which was over a billion more than had been alive a mere 20 years earlier—a stunning surge in population.

Christy's mother describes her as having been a good baby, and remembers how much she enjoyed holding the little girl with the dark hair and eyes. She'd nursed her newborn for six months and taken the baby for all of her prescribed immunizations. She had done everything humanly possible to make Christy healthy. But Christy's wheezing served notice that she had been stricken by a modern plague that would become the central, unwanted feature of her life: asthma.

The American Lung Association defines asthma as a chronic respiratory disease characterized by inflammation of the airways, and increased responsiveness to various stimuli commonly called asthma triggers. Asthma episodes involve progressively worsening shortness of breath, coughing, wheezing or chest tightness, or some combination of these symptoms. The severity of asthma may range from mild to life-threatening.

Asthma is a disease that made its first appearance in the medical records along with the industrial revolution in western Europe. It became more prevalent as our love affair with fossil fuels grew. By 1970, when Christy was born, approximately 3–4 percent of the population of Europe and the United States had been diagnosed with the disease. According to the American Lung Association, among chronic illnesses in children, asthma is the most common. Approximately 33 percent of asthma patients are under the age of 18. The prevalence, morbidity and mortality of asthma in the United States and other countries have been increasing since the early 1970s.[11] For England and Wales, the increase in hospital admissions for asthma for children since the mid-1960s is reported to be sixfold.[12] The U.S. Centers for Disease Control and Prevention reported a 61 percent increase in asthma cases between 1985 and 1995. Asthma is a modern plague.

Numerous scientific studies have established that ozone and sulfur dioxide, both byproducts of burning fossil fuels, are significant factors in the prevalence and severity of the disease. Although these gases are both directly and indirectly detrimental to all of us, for asthmatics they may be life-threatening, since their respiratory systems may shut down when exposed to elevated levels of these gases.

Perhaps because people with asthma make up such a small percentage of

19

the total population, their health issues have not had much impact on reducing our century-old love affair with fossil fuels. True, we have done a great deal to "clean up" our tawdry relationship with oil and coal, but this hasn't eliminated their fundamentally dirty nature. We've even played with nuclear energy on the side, with equally disreputable results. Hydroelectric and solar power weren't always available when we needed them, and we have yet to meet an alternative source of energy we'd want to marry. (Although there are a few, like hydrogen, fuel cells and photovoltaics, that look like good prospects.) Meanwhile, who can blame us for being so smitten with fossil fuels? Abundant, dependable, highly condensed and inexpensive, coal and oil fueled a revolution in lifestyle that we're still enjoying today. For more than a hundred years, fossil fuels have warmed us when were cold, reliably powered our industries and electrical grids, and provided the driving energy behind various forms of locomotion, from the old steam engines to the 100 million automobiles on the roadways of the planet today. Oil has been the source of endless new products, too—plastics, pharmaceuticals, and asphalt, to name a few.

The house Christy lived in was warmed by fossil fuel, the electricity that ran through its wiring was generated by coal, the car that drove her to the doctor was fueled by gasoline. Even her bottles, many of her toys, and her diapers were made with materials produced from fossil fuels. So, even though they contributed to her wheezing, coal and oil were already very much a part of Christy's everyday life and it would be difficult, almost impossible, to imagine removing fossil fuel from the picture.

Of course, at some point we will have to find an alternative to oil. Before the 1970s most of us considered oil to be endlessly abundant. We didn't see oil as a finite commodity. Over the last 140 years, from the time oil first poured out of a well in 1859, we have burned vast quantities of it. It is estimated that the remaining oil will last to at least mid-century at the current rate of usage, based on known reserves.[13] However, estimates may change as new reserves are discovered—such as those found recently in Azerbaijan's Caspian Sea region—and worldwide consumption of oil changes.[14] Oil usage fluctuates in ways that are difficult to anticipate, increased by economic prosperity, population growth, new recovery techniques, and other future variables that we can't anticipate.

So, with such slippery slopes, we must ultimately admit that we really don't know *when* oil will run out, only that it will someday do so. But concerns about the supply of oil may be obscuring a more important issue—as Pieter Tans, a chief scientist at the U.S. National Oceanic and Atmospheric Administration's Carbon Cycle-Greenhouse Gas research group, points out, to burn all our known reserves of oil, natural gas, and, especially, of coal, would be disastrous.

Until there were shortages, it was virtually impossible to see that fossil fuels were at the very root of what we considered to be our way of life in the West—and our way of death. In April 1974, a study of 18 urban areas by the Medical College of Wisconsin showed that urban residents had levels of carbon monoxide (from the burning of fossil fuels) in their blood which were between two and a half and seven times higher than those of people living in the rural United States. At seven times the normal rate, the symptoms of carbon monoxide poisoning were beginning to manifest. The early-winter fuel crisis in 1974, triggered by the oil embargo imposed by OPEC (the Organization of Petroleum-Exporting Countries), led to an unanticipated benefit: a significant drop in the number of deaths in two California counties. The greatest drops in deaths were from heart and chronic lung diseases (including asthma). Lung disease alone dropped 33 percent in San Francisco County and 38 percent in Alameda County.[15] When the fuel crisis ended, the death rates returned to "normal."

Until we were cut off from the endless flow of fossil fuel, few thought about how dependent we'd become financially upon fuel consumption—how entangled were our social, economic and physical needs. Running out of fuel in mid-winter is a catastrophe in much of the world. For Britain, it took just one month of a coal strike, during January 1972, before a state of emergency was declared. It was the first time that over a million British citizens were without jobs. (The only time when unemployment had even approached that level before was during a similar winter fuel crisis in 1947.) Those that remained employed were allowed to work only a three-day week. Ultimately, 1.6 million citizens were unemployed and economic recovery did not begin before electrical consumption was involuntarily cut back 50 percent using a form of rationing. Calling the rationing "staggered blackouts" did not make it any easier for Britons to have their electricity cut off for nine hours every day in the middle of winter. However, their situation was not nearly as bad as it could have been, since without fuel some of the inhabited parts of Earth are as frozen and deadly as Mars.

In China, one of the most environmentally devastated places on Earth today, air pollution, primarily the result of burning coal, is a contributing factor in 25 percent of all deaths (one in four deaths is from lung cancer and cigarette smoking is common, too).[16] The coal burned in China, and the resulting loss of air quality, is undoubtedly a contributing factor in many additional deaths in other countries as well. Air does not respect national boundaries. What is done to the air in China affects the air that all of us breathe, no matter where we live.

Air moves great distances. While it is understandable to most of us that 22–24 percent of the sulphur dioxide found in Korea's atmosphere is from China, many of us don't realize that both dust and pollution are carried across whole oceans.[17] Not only are measurable amounts of carbon monoxide and particulate matter able to travel the 5,000 miles from China to the U.S. in as little as four days, but major dust storms in China have sent enough material in the air to literally turn the sky white over much of the western United States.[18] Californians are breathing East Asian pollution, just as Europeans are breathing pollution from the eastern United States.

Nevertheless, for the Chinese, who have been burning coal since the 1200s, the choice is a relatively easy one. As Chen Qi, a top Chinese environmental official from Liaoning, a region with both a 30 percent unemployment rate and bitterly cold winters, explains: "Heavy pollution may kill you in a hundred days, but without enough heat and food, you die in three."[19]

There is another potential specter in our relationship with fossil fuels that we tend to forget. People are willing to go to war if they feel they are being deprived of what is essential for life: water, land, food, and fuel. It is a problem we have had in the past, as the following story will remind you. It is a problem that we are likely to see again—over other environmental issues—unless we are able to head off the problems associated with global warming.

M ars ruled the skies with its russet brilliance, brighter than the morning star in its glory. In October 1973, Mars, rising with the Pleiades at sunset, warned of a coming war. On the sixth day of that month, during Yom Kippur, the holiest day of the Jewish year, as devout Jews were completing ten days of introspection leading to spiritual reconciliation, violence broke the fragile peace in the Middle East. It was as if, in that most ancient of places, the overpowering crimson light of Mars had awakened a primordial curse: the land and the air forces of Israel, Egypt and Syria battled desperately, while in the Mediterranean, the fleets of the Soviets and the Americans watched and maneuvered. All feared that the war would widen from the cultural heart of Western civilization to its opposing limbs.

I was a physics major then, in college near my home town of Medford, Oregon. In the previous year, after dutifully appearing for my draft physical, I had been designated 1A, putting me at the top of the list when my lottery number was drawn. However, the spreading war in the Holy Land that loomed closer every day promised to be so savage in its violence and so utter in its destruction that one's exact place in line for it seemed irrelevant. As young college students, we simply did not discuss it. We stood in petrified silence as the United States went to DefCon 3, only one quantum level short of full-scale war. To the brink, to the very brink of the abyss we went, because of religion and history and oil.

Then winter came, the oil was cut off by OPEC, and the military equipment at the National Guard armory in our town went from jungle to desert camouflage overnight. The government, perhaps out of a desperate need to create a strategic petroleum reserve for possible war, stopped its ears to its people's cries and allowed the oil companies to hoard all the oil they could. The prices rose astronomically. So, we discovered later, did the profits.

In Oregon, the coldest winter in memory descended upon us. A killing cold from the Canadian tundra lunged straight across the Rocky Mountains and held us in a frigid, prolonged, and deadly embrace. It was an astonishing and freakish occurrence, as if nature, beholding the madness of humanity, had gone mad herself. To a people used to mild, wet winter wind from the Gulf of Alaska and the northern Pacific, the arctic cold that rested on Oregon was disastrous and deadly. Every night the clouds faded, leaving a heartless, clear sky, out of which, it seemed then, all the warmth in the earth fled away into space. The ground, deeply frozen, killed scores of hibernating animals in their too shallow burrows. Whole forests froze, and the city streets were lined with dying trees. Plumbing burst, and desperate homeowners, while trying to thaw pipes with open flames, occasionally ignited their own homes. With heating oil and gasoline in short supply, the deadly cold reaped its toll as the weeks crept by. The intoxicated and careless died in their cars; the old, poor and hapless in their homes.

As we sat in the frigid, mile-long line of cars waiting for what little fuel was available, my girlfriend, Faye, and I held each other close for warmth. I could suddenly see what had been invisible to me before. The life of my town, my country, my whole way of life depended on oil, and people were dying—from cold, from war. Oil was killing us. Expressing my dismay to Faye, I swore to God that I would one day free humanity from its mortal addiction to the viscous black substance we pumped from the ground. Someday, I would do battle with oil. We sat huddled together in the frozen landscape, in that coldest winter of the Cold War, creeping forward car length by car length, while overhead, in the clear winter sky, the glow of Mars began to fade.

So we have begun our tale of two planets, one dead, another dying—a story of life and death on Mars and Earth. What lies ahead is as essential as breathing, as remarkable as a visitor arriving on our threshold from another world, and as disturbing as a house on fire—especially when it's our own. Welcome to life in the cosmos.

CHAPTER 2

The Airless Sphere

I n the depths of space, on the ecliptic plane,[1] 135 million miles from Earth, the stars shone brightly, like sparkling, blue dust. If you'd listened to the radio spectrum you'd have heard only the faint whisper of the void. It was late in the day on June 19, 1976.

Suddenly, with a crashing whine of radio transmission, a sparkling, blue and gold spacecraft flashed past at five miles a second, streaking relentlessly toward the somber red planet Mars, which loomed in the starry distance. Packets of digital radio data filled the ether with a humming chatter in the spacecraft's wake, as messages poured back and forth between the spacecraft and the distant Earth.

As it neared Mars, the spacecraft wheeled elegantly to point its rocket engines toward the cloud-flecked planet. Then, at exactly the right moment, as it circled over the gleaming northern polar cap, the spacecraft fired its engines. Cones of blue flame poured out for several minutes, slowing the spacecraft until it dropped deeply into the gravitational embrace of Mars and finished its bullet-like rush across the inner solar system by falling into orbit around the red planet. When the orbit had stabilized, the craft rolled to direct its camera to Mars, and its gleaming blue solar panels sprouted in four directions from a hexagonal body sheathed in shining gold. Beneath the body of the spacecraft hung a domed vessel carrying a landing craft.

Cheers broke out in the control room at the Jet Propulsion Laboratory (JPL) as the spacecraft radioed back its first picture of the planet's surface and confirmed that it had achieved orbit.

Viking A had been sent to Mars to probe for signs of past water flows and volcanic eruptions, but, far more importantly, to seek signs of microbial life. Its landing craft, VL-1, carried ingeniously constructed life experiments for testing the soil at the landing site. The detection of life on the planet Mars would be the discovery of the century, perhaps of the millennium. It would move humanity forever beyond its world view of the Earth as the only world that counted, with the rest of the cosmos as merely a curious distraction. We stood on the threshold of new discovery, a moment of both excitement and peril, when tantalizing possibilities both promised and threatened to be revealed. However, as with even the best-laid plans, this unfolding drama was soon accompanied by frustration.

Mars was not following the plans laid out for it on Earth. The proposed landing of VL-1 was to be in Chryse, near the mouth of the Valles Marineris canyon system, close to the Martian equator. But the cameras of the Viking A were already showing that the proposed landing site was very rocky. If the lander hit a large rock on the way down, it could flip over and be destroyed. The plans had called for a landing of VL-1 to occur on July 4, 1976, the day of the American Independence Bicentennial, but this plan was now cast aside.

The Viking cameras were looking at a Martian surface with unprecedented magnification, a factor of five greater than those of Mariner 9, which had preceded Viking in 1971. The pictures obtained from the trusty Mariner 9 had been the basis for the choice of landing sites for the VL-1 and its sister VL-2, which was even then streaking toward Mars aboard Viking B to join Viking A in orbit. The VL-2 was aimed at the prime landing site, an area to the north of Chryse called Cydonia, where it was thought life and water were most likely to be found.

Temperatures and atmospheric conditions on Mars generally resemble those of high desert plateaus in the terrestrial Antarctic. However, the Cydonia region is exceptional in that both temperatures and the abundance of water vapor are high relative to the rest of Mars. A low plain runs through Cydonia and stretches north and south, so that in the Martian summer masses of water vapor from the frigid northern polar cap have an unimpeded path south to the warmer temperate regions, where temperatures often rise above freezing. Cydonia is a place where both water and temperatures above freezing could come together periodically. If life hangs on grimly anywhere on Mars, perhaps as a dormant remnant from some earlier wet and warmer period, it is most likely to be in Cydonia. In addition to the possibility of water, Cydonia is a region of transition between the rugged uplands and the low, smooth northern plains of Mars, exposing a varied terrain of interest to geologists as well as biologists. Therefore, Cydonia was picked as the prime landing site for the VL-2, but the VL-2 landing was scheduled to take place only after the supposedly less risky landing of VL-1 at Chryse.

Chryse is where three vast water channels from the Valles Marineris and nearby regions had emptied in eons past. Not surprisingly, it looks like a dried-up seabed. A few scientists believed there was a reasonable chance of finding living microbes there, perhaps waiting patiently for water to return.

As Viking A's cameras, performing flawlessly, were imaging Chryse and Cydonia, looking for a safe place to land, powerful radar beams from Earth were sweeping Mars, probing for a place without large boulders. In Chryse, a smoother spot was found to the west of the original site. When the new landing location was certified by the pensive board of scientists at JPL, the signal was sent and the siblings, Vikings A and B, obeyed the commands flashed across the starry gulf toward Mars.

Viking A rode in high orbit around Mars, performing almost perfectly, while its sister craft, only weeks away from arrival, was also on its best behavior. It was apparent that the United States, after long years of waiting and planning, had finally arrived at Mars in force. Humanity would, for the first time, be able to answer the great question: Did Mars once live?

The image of planet Mars as alive had for so long been a fixation of the scientific and popular consciousness that it had become embedded in human culture. Mars was a name that evoked myriad images—from beautiful princesses to alien invasions. Mars was an everyday synonym for "impossibly remote strangeness." "He's from Mars," one would say of a peculiar colleague, or "The Martian" would be invoked to explain an otherwise inexplicable mishap. The popular image—promoted heavily by Mars observer and astronomical writer, Percival Lowell (who depicted Mars as a world uncannily like the desert of Arizona and capable of sustaining life despite its thin air),[2] and science fiction writer, Edgar Rice Burroughs—was that Mars was populated. Based on popular culture, the expectation of some people was that Mars would be like Earth, except that on Mars people didn't wear clothes. However, this popular image was greatly at odds with the thinking of the scientists on the Viking team.

The arrival of their spacecraft at Mars caused a curious mixture of elation and depression among the members of the Viking science team. On the one hand, the prospect of massive amounts of new data about a fascinating other-world was cause for elation. On the other hand, theirs was a fool's errand. The opinion most of them shared was that going to look for life in the Martian soil was a doomed enterprise, certain to give negative results while raising expectations in the public's perception. As one project scientist put it in an address to assembled scientists at Lawrence Livermore National Laboratory, the main motivation for the life experiments came more from Lowell's book on the Martian canals in the 1920s than any modern understanding of Mars. The Viking life experiments were, in this scientist's opinion, "a waste of space on the lander." The prevailing opinion on the Viking science team was that Mars was lifeless and had always been lifeless, like the Moon. Looking for life on Mars was an exercise in futility.

This view, which we choose to call the Lunar Mars theory, was curiously at odds with the evidence available about Mars in 1976. Data had been gathered for several years by Mariner 9, a spacecraft that had first orbited Mars in 1971. Mariner 9 had completely mapped Mars' surface to kilometer scale, and in isolated places at much higher resolution. It had also discovered the vast Valles Marineris canyon system at the equator, with water channels, now dry, emanating from its mouth.

In general, Mariner 9 had discovered a new Mars. It had found that the northern hemisphere was quite young, covered with low, smooth plains, and reworked recently in geologic time by what appeared to be flowing water. This was completely unexpected. Based on the images of the southern portions of Mars obtained on previous missions, Mars had appeared dry, barren, and Moon-like. This new information meant that Mars, for some period of time, had experienced Earth-like temperatures and atmospheric pressure, both essential for the presence of liquid water. In the rugged highlands of the southern regions of Mars, Mariner 9 had confirmed that the landscape was heavily cratered—indeed, like the highlands of the Moon. However, these highlands were criss-crossed with distinctly unlunar water channels. Even now it is difficult for us to take in the implications of this, to accept that our nearest planetary neighbor—that frozen, airless sphere next stop out from the Sun—at one time had water. Like Earth, Mars was a dynamic planet with recently active volcanoes and a warm and wet past.

However, it was the lunar quality to the southern highlands of Mars that had fixed the consensus within the Mars science community. Those highlands were so similar to those found on the Moon in their extent of cratering and apparent age that it appeared Mars could not have had an Earth-like climate, and thus no significant erosion, for nearly 4.5 billion years. Added to this was the harshness of Mars' surface environment. It has only 0.5 percent of Earth's atmospheric pressure, an atmosphere of mostly carbon dioxide, no ozone layer to stop the deadly rain of ultraviolet rays that would destroy all exposed Earth-like life, and temperatures far below freezing during most of the year.

Together the Moon-like southern highlands and the harsh surface conditions gave rise to the Lunar Mars concept: Mars was dead. Both geologically and biologically, it seemed obvious that, like the Earth's Moon, Mars had been dead for all or almost all its history. A few lonely voices were raised to argue that life could have arisen on Mars in its early warm and wet period, perhaps in the numerous water channels amid the craters, and that a few of its hardiest microbes could have survived until the cold, dry present. However, some scientists, being by nature skeptical and independent, questioned whether the channels were even caused by water, speculating that some exotic fluids could be flowing freely under cold, dry conditions. So, despite the Mariner 9 pictures, the consensus of the Viking team had essentially been established in the 1960s, over a decade earlier; in 1976, cynicism about the possibility of life on Mars still prevailed.

And then there was the question of funding: if Mars had life, we'd have to send people to explore it. If we sent people, Johnson Space Center in Houston and NASA proper would get the lion's share of the funding. If Mars was dead, much more of the money would go to JPL. So the search for life on Mars also offered a huge payout—and terrible financial losses for the loser.

But there was a time in our history—not so long ago—when science was not about guarding one's source of funding. There *was* no source of funding.

29

Science was its own reward. Enthusiasts most probably earned a living doing something other than science and followed scientific pursuits in every moment that they could liberate for that purpose. They did so because they were deeply curious about the nature of reality. There was a time when the mind of a scientist wasn't just open, it was a veritable font of fresh and powerful questions that led to an unfolding of understanding. Science wasn't a detached professional pursuit. It wasn't just about technology or specialization. It was about questions—questions that produced an insistent force that grew inside the individual, encouraging him or her to push against the limits, questioning even the obvious, the apparent, the assumed. The domain of science was populated by gifted, driven amateurs who found the inquiry into the nature of the universe as compelling as life itself. There once was a time when the driving force of science was "amare," or love—a burning, passionate love of inquiry.

It might have caused a fair amount of gossip when a Unitarian Minister named Joseph Priestley wandered from his home to the brewery next door in pursuit of a substance that was produced there in large wooden vats. But when Priestley returned home perfectly sober, carrying glass bottles that appeared to be empty, one can only assume that the gossip increased rather than subsided. "What," his neighbors must have wondered, "can he be up to *this* time?"

This was not the first time, nor would it the last, that Priestley's neighbors in England of the late 1760s were baffled and dismayed by the man who had taken over a pastorate in the town of Leeds. Joseph Priestley was an individual who had decided opinions and a reputation for a hot temper. Worse yet, his strongly expressed religious and political beliefs differed wildly from those shared by most locals. Nevertheless, his frequent trips to the brewery were scientific expeditions, not religious pilgrimages, for it was in the brewery that he was able to obtain plentiful supplies of "fixed air," which bubbled up to the top of the vats and which he could use in the many experiments that were part of his compelling avocation: the study of atmospheric gases and respiration.

"Fixed air" had been isolated just a few years earlier by a Scotsman named Joseph Black, who confirmed its lethal nature when a bird he'd placed in a jar filled with the substance died within ten seconds. Joseph Black also observed that when he breathed repeatedly into a bag, his lungs would eventually feel like they could no longer expand. Black reasoned that the substance that normally allowed his lungs to be "elastic" was gone from the air after a number of breaths, and that was probably why the bird had died. Although his understanding was inaccurate, the phenomena Black witnessed with the bird and with his own lungs were related. Both were caused by the substance left in the bag.

Joseph Priestley, intrigued by Black's new, invisible substance, found supplies of his own to experiment with; thus his trips to the brewery to retrieve bottles of the gas that collected at the top of the vats. Among other discoveries, Priestley found that when this lethal "fixed gas" was collected under water, some of the gas dissolved in the water, causing bubbles to form and giving it a characteristic tartness. We don't know whose idea it was to taste the bubbly water, filled as it was with this gaseous substance that behaved with such toxicity, but obviously scientists back then were sometimes their own guinea pigs. If Priestley hadn't survived that experiment, then a whole industry based on the bubbly water might never have been born.

Had he kept to his experiments with gases, Joseph Priestley might have stayed out of trouble. But the same driving passion that ignited his scientific pursuits also fueled his support of the French revolutionaries, which didn't make him popular with the King of England, and an angry mob burned Priestley's home, chapel, and laboratory. However, Priestley seems to have taken the hint and first moved to London, then emigrated to the U.S., where he founded an early Unitarian church in his town and where ultimately his discoveries would be appreciated by many people far into the future in ways he'd never intended. In fact, you can enjoy Priestley's discovery today just by drinking a can of soda pop.

Although many of us today consider science a pursuit of cold, hard fact, it has always been and continues to be a human endeavor where personalities, prestige, power, and now, certainly, funding, emerge as factors in scientific argument. Scientific disputes can quickly become personal and political, and the question of life on Mars was no exception. Since the consensus held another view, those who viewed Mars as having been more Earth-like in its past worried that they might be perceived as dreamers or "believers" rather than true scientists. The scientific mainstream, slow to embrace new ideas under the best of circumstances, was terrified of being found investigating anything that might bring ridicule or embarrassment. Life on Mars as a serious scientific inquiry was exactly the kind of pursuit that could—and did—expose the investigators to professional ridicule, or worse. Anyone who thinks that the investigation of Mars has been purely a scientific pursuit is not familiar with the denial, misrepresentations, and obscuring of data that have been the hallmarks of an unwanted shift in scientific reality. Life on Mars was as welcomed by most of the scientists at JPL as Joseph Priestley's support of the French Revolution was by the King of England—and for much the same reason. Therefore, the Viking's quest for life on Mars was cause for ambivalence among the project leaders and contention among the staff scientists, in part because it conflicted with what they saw as the realities of Mars' physical environment and geological history, and in part because of potential embarrassment and disruption of funding. There existed however, a deeper reason for their ambivalence. It was rarely spoken of, yet it was real and tangible, and it was not confined to scientists. To find life, even primitive life on Mars, past or present, would demote humanity. Life on Mars would not only revolutionize biology, astronomy, planetary science and countless other scientific disciplines, it would radically—and for some people, disturbingly— alter world-views. Very simply, we feared being pushed in an instant from being the rulers of the cosmos to being rulers of a speck of dust.

The truth is, humanity rules Earth in splendid isolation from the cosmos. Since we perceive our world as that portion of the universe where all things of any "importance" occur, we sit like rulers of an oasis in the center of the Sahara, masters of all that matters. We annually crown a new Miss Universe with not even a wise smile that this is an absurd joke. Far from it, Miss Universe is serious business. We not only don't give a thought to the rest of the universe—as far as humans are concerned, we *are* the universe. If we found microbes on Mars, the most likely place for them in our solar system, then why couldn't there be life—even intelligent life—elsewhere in the cosmos? For many of us, the night sky would hold a new-found wonder. But for many others, it would be ablaze with potential threats from hostile new worlds.

In general, the United States in 1976 wasn't anxious to find life on Mars. After a decade of turmoil—including the Vietnam war, the protests, the U.S. pullout and the final collapse, Watergate, the Yom Kippur war in the Middle East and the oil embargo—something in the public mind cried out for quietness and reassurance.

Perhaps that is why, when six scientists appeared before the United States Congress to warn the world of the destruction of the ozone layer by substances known as CFCs (chlorofluorocarbons), including a commercial form of the gas produced by DuPont and Co., nothing was done.[3] A spokesperson for DuPont called the ozone research "purely speculative," claiming that restrictive legislation was "unwarranted at this time." The Consumer Products Safety Commission agreed with that opinion when they also rejected a proposed ban on Freon-based spray cans because they felt there was insufficient evidence to support a ban that year. Despite official inaction, many consumers took matters into their own hands by boycotting aerosol products (which used CFC as a propellant gas) long before governmental agencies and corporations acted to protect the planet's atmosphere. The decision to delay action on CFCs has conservatively cost humanity at least 6,000 cases of skin cancer and death every year for the past 20 years.[4] Of course, CFCs ultimately were regulated. However, the time lost by failing to act with velocity will cause countless numbers of cancers and deaths every year for decades and decades to come. This is not an abstraction: the failure to act on behalf of planetary concerns has real, direct costs to humans. The people who will bear the consequences of this failure will be people that you know, members of your family, and possibly even you. (For more information on CFCs, *see* pages 57–8.)

Fortunately, Mars exploration held no such threat for mankind. While conclusions about life on Mars had been prematurely drawn by a group of scientists who should have been more open to that possibility, the outcome would not be lethal to humans on Earth. As the day of the landing in Chryse approached, the United States prepared to go boldly where no human had gone before, but it was not looking for trouble. It was looking for reassurance. Sour complaints about wasting space for life experiments on the Viking lander had a comforting side, as cynicism often does. They assured everyone concerned that the life experiments would most certainly have a negative result. Everyone, from the stock market portfolio manager concerned about the reaction on Wall Street to the stunning Miss Universe 1976, could rest easy. The boat humanity rode on was not going to be rocked. Mars was, after all, like the Moon, and the life experiments held inside the gleaming spacecraft were most likely useless baggage.

Nevertheless, even as Viking A orbited and prepared for the landing in Chryse, it continued to take swath after swath of images at high resolution (by the standards of the day), scouring the northern plains of Mars looking for a new landing site for VL-2. The rest of the world held its breath. In a matter of hours we would be exploring another world. This would be the first time we would land on another planet. Whether we would find life there or not, this much everyone could agree on: from what we knew about Mars, it seemed to be a far more hostile world than our own.

Above the bright aura of Mars, against the backdrop of its wispy atmosphere, on July 20, 1976, the Viking lander vessel detached from Viking A and began its fall into the thin atmosphere of Mars. Glowing like a shooting star, the domed lower vessel, which had been released with an explosive blast of fire, plunged flaming into the Martian ionosphere where the atmospheric resistance began to slow its manic descent. Inside the shell of this transport craft was the tiny lander VL-1, heading toward the red plain of Chryse, the "field of gold," where in eons past vast torrents of water had emptied. At the correct altitude, the lander shell expelled and deployed a parachute. Finally, 650 feet above the surface of Mars, the domed vessel blew open and disgorged the VL-1, which ignited its own rockets and floated down gently on pillars of fire to the Martian surface. Like the Apollo 11 landing on our Moon in 1969, it was an astonishing achievement. Never before had a successful landing been made on another planet; never before had technology triumphed so brilliantly so far away from home.

So, there we were, poised at the threshold of another planet's doorway for the very first time in human history, finally ready, after waiting all of eternity (including the sum total of the period required for the evolution of our species to the point to where we could actually take that monumental step). But what next? Whatever followed was likely to become legendary because that moment had been anticipated in the imagination of humans for thousands of years. It was a moment that veritably screamed: Lights! Roll 'em! Action! However, there were no humans on board to enact the drama as there had been when Neil Armstrong stepped onto the surface of the Moon for the first time. But there were cameras, and if a great science fiction writer like Jules Verne had been writing the story, the action might have gone something like this:

Pshhhhhh . . . Air rushed through the mouth valve on the front of the steel helmet one last time before he turned the thumbscrew to close the valve and sealed off his respiration from the outside world, a world that he could now only view through a round window of heavy mica reinforced by a grid of steel rods. Pressed to his face by an inflated pneumatic washer (which encircled his head forming a seal between flesh and steel), the helmet was all that would keep him alive as he explored. His final exhalation before the valve shut had filled two leather bladders that hung down from the front of the helmet like lungs protected by a leather apron.

From the moment the valve was closed, normal respiration ceased. The air the helmeted man breathed circulated through a series of heavy canisters he wore strapped to his back in a 40-pound knapsack. At the bottom, two large canisters were filled with oxygen. Above those, two smaller cartridges were filled with loose crystals of potassium hydrate which absorbed carbon dioxide. Connected by a grotesque series of tubes, this system was designed to remove waste from the wearer's breath and to recirculate the nitrogen atoms by temporarily saving them in the leather lungs, returning them to the human's lungs only after mixing them with fresh oxygen from the canisters. Thus, he would be able to survive in a hellish, alien environment.[5]

Joining his crew, all dressed in similar gear, the helmeted man could see a clock on the back of each crewman's knapsack, as they could see his. When the group prepared to descend, the analog clock dials were reset to count down from two hours to zero in quarter-hour increments—counting off all the time they would have to explore the deadly terrain beneath their feet. After a moment's apprehension, they began to descend.

The temperatures, although still uncomfortably high, had cooled considerably from the red-hot heat of a few weeks earlier when there had been a massive fire fueled by the carbonaceous substance that ran in dark veins underground. Still remaining were the toxic gases that had filled the enclosure, and they were capable of killing within seconds. So, when this crew returned to the site of the fire to find and retrieve the dead bodies of those who had succumbed in the smoke-filled catacombs, they had to prepare for the now noxious environment as if for travel on the surface of some distant planet.

Wait a minute . . . This is the wrong movie. This is not some nightmarish pseudo Jules Verne version of the arrival of humans on Mars. This is a very real account of an event that occurred on Earth. These helmet-wearing men were coal miners in the early 1900s and where they were going, all suited up in leather and mica, was into the Earth itself. The deadly gases that filled the mine and threatened to extinguish their lives were what they, like miners everywhere, called whitedamp, blackdamp, and the terrible mixture of the two that remained after an underground fire: after-damp.

Whitedamp is the tasteless, colorless, odorless gas we know as carbon monoxide, which sometimes combines with the highly explosive gas, methane, producing a mixture known as firedamp. Blackdamp is the tasteless, colorless gas that miners in the early 1900s knew as the deadly "carbonic acid gas." They also called it chokedamp because of its ability to choke those who inhaled it. Miners treated blackdamp with as much respect as a substance with such lethal potential deserved. Carbonic acid gas could—and frequently did—kill miners.

By the beginning of the 20th century, all miners wore or carried special lamps that would flare in the presence of explosive methane gas. However, before monitoring equipment was developed, in order to detect the invisible non-explosive gases like carbon monoxide and carbon dioxide, crews of miners frequently carried a canary in a cage down the dark shafts into the mines. Canaries, which have more sensitive systems than humans, would stop singing, fall off their perches and lie gasping until dead at the bottom of their cages—just like the birds in Joseph Black's jar—if exposed to even a little too much carbonic acid gas. A silent canary was a signal to leave the mine immediately—if you were able to.

Christy Booker was as excited as a little bird, and it wasn't because a spacecraft was going to Mars. She was on her way to kindergarten. While the Vikings were powering through space on their way to Mars, Christy was going to school for the first time. The photos her mother took of her dressed in her new school clothes showed a little girl who looked like a healthy, normal, active five-year-old, and most of the time she was. As long as she didn't run too hard or inadvertently come into contact with a whole list of allergens that triggered an attack of asthma, she was just fine.

At about the age of five, Christy began to take a series of shots administered by her allergist, one of the two physicians who treated her routinely. The shots were the source of a great deal of dreaded anticipation. To a young child, a hypodermic needle loomed large, dripping like a horror movie close-up. Getting "shot" was a painful and fearful experience. Yet once each week, for week after week, year after year, Christy dutifully had her allergy shots, driven like a lamb to the slaughter by her insistent mother to see a persistent needle-wielding doctor. All of these efforts were made in the hope of keeping her asthma in check, simply to help Christy breathe.

Of course, when Christy had a cold, it was a more serious matter than it was for a non-asthmatic child. The average child between four and fifteen years old has between six and ten colds per year. So, like most asthmatic children, Christy spent more of her childhood at home recovering from the colds she inevitably caught. (In the U.S., asthma is the number one cause of school absences attributed to chronic conditions.)[6] It was harder for Christy to breathe when she had a cold. In a recent interview she said it felt like breathing took extra concentration, a real conscious effort just to inhale and exhale. Her respiratory system labored harder with the stress of an infection, which increased the chances of a full-blown respiratory crisis, and no matter how mild the cold, full recovery took more time. According to the American Lung Association, one study estimated that in the U.S. in 1994, school days lost to asthma amounted to $673.2 million in parents' and other care givers' time lost from work.

Several times in her early childhood Christy was brought to her pediatrician's office in a respiratory crisis, the kind of emergency that would routinely be treated in a hospital emergency room today. In the early 1970s, a common treatment for children in an asthma-induced respiratory crisis was the administration of the hormone adrenaline, which was injected into the bloodstream with a hypodermic needle. Christy's tiny system, powerfully stimulated by the "fight or flight" hormone, would respond immediately. Physiologically, her body responded as if she were being attacked. She said she could feel her heart pounding frantically in her chest and her tiny musculature surging, like a marathon runner's at the starting line. Psychologically, she was thrust into a highly anxious, agitated state—on top of the stress of not being able to breathe very well. As the powerful agent washed through her system, she was unable to sit still any longer. Back and forth, back and forth, she madly paced in the tiny medical office until her artificially overstimulated system somehow eased her breathing just enough so that her doctor could administer another medication that would act as an antidote for the first, easing her panic, but leaving her weak and exhausted.

So, Christy was like every other child in her kindergarten class except for one thing: they could reliably take the next breath and the next. Christy could most of the time, but she routinely had to work a lot harder and regularly face discomfort to be able to do so, and sometimes, the process of taking her next breath turned into a crisis that involved her whole body, her family members, her doctor, and a medical intervention that would probably frighten the bravest of us.

As the shouts of joy and triumph in the JPL control room ebbed, the first picture from the surface of Mars appeared on the screens: a beautiful close-up of the lander footpad resting securely on the soil of Mars. The control room personnel held their breaths as the camera panned upward to the sky and a magnificent panorama unfolded in black and white on the great projection screens before them. There were gasps of awe at the beauty and austerity of the scene. The first image of the Martian landscape, in crisp focus, showed a desert studded with boulders stretching to the far horizon beneath a clear sky. Everyone was astonished that the craft had managed to land safely in such a field of rubble.

On Earth, news of the American triumph swept the world. Even in the Kremlin there were smiles and happy nods. In north-central Europe, where steeled armies of millions faced each other across the West German frontier, the Warsaw Pact mentally re-adjusted its assessment of NATO strength upward, in view of the fact that America had once again done something—like the Apollo Moon landings—that the Soviets could not do.

The following day was equally auspicious for JPL when the first color images were received. As a crowd surrounded the monitors, the colors of the Martian landscape displayed themselves like an Arizona picture postcard: bright sunlight, blue sky flecked with hints of high clouds, reddish-brown soil and gray rocks. The technicians, scientists and media representatives were spellbound. It looked just like home—at least it looked very familiar to anyone who had see the desert regions in the southwest of the United States. Who would have imagined that our first look at another world would appear so Earthly? An American flag was proudly emblazoned on a domed compartment of the spacecraft which figured prominently in the scene. Never had its red, white, and blue colors glowed so brightly as they did beneath the Martian sky that day. For the first time humans beheld the sky of a planet different from their own—and it was a reassuring deep blue.

Amid the celebration, a very perplexed, newly graduated high school student named Ron Levin sat at a console at JPL. For over two hours Ron watched with the others as color images came in. Ron, the son of Dr. Gilbert Levin, a scientist connected with one of the tests aboard the lander, was fascinated. Like the rest of us, Ron saw the "blue sky" of Mars in the first color image. Unlike us, Ron also saw subsequent images of rocks with greenish patches on them. While he watched, a Viking imaging team technician adjusted the color controls on every monitor in the laboratory, increasing the red channel so that the sky lost its blue tint and the color of all of the rocks shifted from an earthy reddish-brown to an unearthly, deep rusty-red hue. When Ron attempted to adjust the monitors back to the original settings, he was told he'd be ejected from JPL if he persisted. To this day, Ron, now Dr. Ron Levin, physicist, makes it clear that, in his considerable estimation, there was no scientific justification for the adjustment in color.[7] He speculates that the color was changed primarily because the planetary scientists took a dim view of the greenish patches on the rocks, which might have suggested to the public that some primitive form of plant life might be growing right on the surface of Mars. The blue sky of Mars has been red ever since.[8, 9]

While JPL image processing technicians fiddled with Mars' colors, the VL-1 tackled the question that was on everybody's mind—was there life on Mars? A robotic shovel scooped up a small sample of soil and deposited it in a receptacle on the lander. Soon the world would know whether life, perhaps merely long-dormant spores, existed in the soil of Mars. The tests were elegantly designed and based on several different principles to ensure that the net for possible biological activity on Mars was cast wide. They included the Pyrolitic Release experiment (PR), the Labeled Release experiment (LR), and the Gas Exchange experiment (GEx). The first two used nutrients labeled with radioactive carbon-14 to test whether agents in the Martian soil acted as if they had a metabolism; that is, if they processed the nutrients or gases into bound states or through respiration. The Gas Exchange experiment also looked for exhaled gases, but without radioactive labels. But of course it was a "fact" that Mars was dead and so these experiments were "obviously" just an exercise in brilliant futility.

However, immediately it became apparent that Mars was not giving the answers scripted for it. Its soil released free oxygen on being moistened with water. As in the plot of *Robinson Crusoe on Mars*, a 1960s science fiction movie where an astronaut shipwrecked on Mars saved his own life by discovering oxygen in the soil, the soil that was sampled from Chryse was full of oxygen and very chemically active, and this threw the biologists into a panic. It meant that interpretation of the results of the biological experiments, which were also looking for chemical activity, would be very complicated.

They immediately found a great deal of exactly what they were testing for: chemical activity. This also meant that the positive results of the life tests were ambiguous, a potentially nightmarish circumstance for the scientists involved. Almost immediately a major scientific stew over the meaning of the results of the life tests began in public.

While life experiments were being performed at Chryse and arguments about the life tests raged on Earth, Viking A was still orbiting, taking swaths of pictures as part of the search for an alternative landing site in Cydonia. Five days after the landing in Chryse, it was completing its survey. Without so much as a blink of its mechanical eye, the camera systematically scanned the Martian surface, and recording the information digitally, it took a picture of something curious.

The curious picture was discovered by Tobias Owen, a member of the NASA imaging team, while searching the Mars landscape with a magnifying glass for the perfect landing place. On frame 35A72, the 72nd frame taken on orbit 35 of the A spacecraft, something astonishing appeared, something that looked like a "face," a humanoid face that seemed to be carved into a mesa nearly a mile across. The image was immediately shown to Harold Masursky, a chief mission scientist, who released the picture at the daily press briefing for the Viking mission. It caused an immediate sensation.

However, the strange "face" appeared during the middle of the controversy over the life detection test results. Both of these messy public controversies about "life on Mars" were annoying distractions from the mission to a lunar Mars the scientists felt they were managing. So, it had to be put to bed quickly. It was therefore announced to the press corps that Viking A had taken another image a few hours after the first, and in this image the "face" had disappeared, so it was obviously a trick of "light and shadow." The image 35A72 had been taken at local sunset in Cydonia, and thus any image afterward would have shown only darkness.[10] But no one in the press room could have known this, so the second announcement was accepted without question. Although this misrepresentation would fuel conspiracy theories for decades, the immediate problem was temporarily fixed. The second announcement gave reassuring closure to the Cydonia matter, even if it was not accurate.

The pressroom resumed its focus on the life detection test results, where a tense situation was emerging. Based on rules adopted before the landing, two of the three life experiments were giving positive results, and that meant a positive life verdict. In the councils of the Viking leadership, Dr. Gilbert Levin, the scientist who had developed the Labeled Release (LR) test, one of the two tests that had given positive results, and Robert Jastrow, a team member who supported the life findings, argued passionately that life had indeed been detected. However, to the Lunar Mars faction, this was a potentially disastrous finding. The Lunar Mars proponents scrambled for a way to recover the initiative.

In order to fix this "life" problem, another instrument was brought into the life determination experiment. This was the gas chromatograph, which did not look for life but rather the organic chemicals necessary for life. It showed clearly that Mars' soil was very sterile, more sterile than the Moon's, where organic matter from meteorites impacting over many eons could be detected. In fact, the gas chromatograph's test results supported the conclusion that something like hydrogen peroxide was present in the Martian soil. It therefore appeared that Mars' soil was not only bare of detectable levels of organics, but was positively hostile to organic matter, attacking it chemically like bleach. The problem was solved. The soil was full of antiseptic. It could hold no life.

That it was possible for a few hardy bacterial spores in a sample to be totally missed by the gas chromatograph (as had happened in a similar test of a soil sample from Antarctica) and still give a positive life response in the experiments was swept aside.[11] The verdict of no life was satisfying to many on the Viking project. The gas chromatograph had evened the score, making "no life" a reasonable interpretation of the results, and it gave closure to the investigation: Mars was like the Moon, just as most of the scientists had expected.

On August 2, 1976, the Viking B spacecraft assumed Mars orbit. The only drama now remaining was the still unresolved question of where to land the VL-2 to ensure maximum scientific return. The landing site in Chryse had been good, but the observers on Earth had been unnerved by the presence of a large boulder near the lander that they named Big Joe. If such a boulder was beneath the lander when it touched down, it would destroy it. Radar sweeps from Earth finally identified a smooth place that appeared free of boulders. Though the landing site obviously lay in the debris field of a large crater named Mie, the radar results argued that the site was sure to be covered with many feet of sand. The final decision was made to land near Mie in the Utopia region.

The long-awaited landing was a success: when the VL-2 cameras took a panoramic scan of the surroundings the merriment in the control room turned to astonishment and chagrin. The lander had miraculously touched down safely in a field of large boulders stretching to the horizon in every direction. Later it was found that the radar returns on which the final decision had been based were from the empty sky. The radar antenna had jammed and was no longer pointed at Mars. So, as with most utopias, this one had looked good only from a distance.

As the months crept by, the two landers on Mars sent images of the Martian sunrise and sunset, dust storms and morning frost. They even recorded a total eclipse by Phobos, one of Mars' moons. Meanwhile, life on Earth continued. Miss Universe 1976 completed her reign. The "Mars as the Moon" concept was secure in the public's perception, and so was the human sense of self-importance. The sticky details of the Mars life experiments, the question of a blue sky, and every other anomaly faded to a warm sense of accomplishment in the Viking team—except for Gil Levin, who now was more convinced than ever that the positive results from his life experiment meant just what they appeared to mean.

The Viking landers operated for several years before communications were lost. All in all, the Viking orbiters and landers had been spectacularly successful. We had gone to Mars in force, boldly asked the big questions, and had received answers. The success of the Viking expedition is a testament to the leadership and talent of the Jet Propulsion Laboratory and to the vision and financial investment of the United States as a whole. However, the leaders of the Viking expedition had also left their own philosophical stamp on the preliminary interpretation of the data: that Mars was Moon-like as opposed to Earth-like in history, that it was lifeless and had probably always been lifeless. They had brought the Mars problem to some sort of conclusion, and chased away the threatening specter: the microbes of Mars, and the threat they posed to humanity's sense of security. The Vikings generated a mass of data so vast it would take decades to realize its full importance. But, slowly, these data began to tell a different story than the preliminary analysis had shown. Loose ends of the story began to unravel with time. The Lunar Mars concept slowly came under siege.

Mars was not the only place where important planetary data was being collected. By the time the Vikings were launched in 1975, atmospheric gases had been monitored on Earth for more than two decades. One of the most important collection points was at the summit of a volcano on an island in the middle of the Pacific Ocean. The Hawaiian site for the Mauna Loa Observatory was selected because of its geographic remoteness. Over 2,000 miles from the nearest major landmass, the observatory was constructed at an elevation of 13,455 feet above sea level.

Originally a weather station, the first structure that the U.S. Weather Bureau built on Mauna Loa (with the cooperation of the Territorial Department of Institutions) was very small, a mere shack of 8 by 10 feet.[12] In order to reach the summit, one had to drive up a cinder road to an elevation of 9,300 feet. From there, the trek to the top had to be completed on an exceedingly steep, rock-strewn course in a four-wheel-drive vehicle. Paradoxically, bad weather and occasional snow made the rocky trip to the Hawaiian summit even more dangerous.

The marine air that arrives in Hawaii on the north-east trade winds has generally not passed over land for many weeks. So, the air that is tested at Mauna Loa is a representative indicator of the cleanest air available anywhere on Earth. The air is collected from several locations near the summit and is evaluated periodically from samples captured in flasks. It is also continuously measured from samples fed to electronic measuring devices via feed lines to the intake ducts. Every effort is made to ensure that the air samples are not tampered with. The collection sites are secured by fences, and air pumps continuously flush the lines that deliver air samples to the sensitive monitoring devices. These elaborate precautions are taken in order to assure the purity of the "benchmark" that is continuously being established. The air measurements at Mauna Loa have always been a standard to which all other atmospheric readings are compared, so every possible precaution is taken to ensure that nothing in the local environment changes the quality of the "global" air measurement.

There was just one practical problem with this theoretical purity. Every time a vehicle labored up the side of the mountain carrying a new crew or supplies to the station, "huge spikes" caused by the carbon dioxide and particulate matter the car released would appear on the paper printouts from the electronic equipment that was continuously measuring these substances.[13] Thus, even on the wide-open, treeless, windblown slopes of a remote volcanic mountain, the exhaust of a single automobile was sufficient to change the mixture of the "global" air being sampled.

"It's our effort to protect our environment up there—to keep us alive," Dr. Pueschel, director of the Mauna Loa Observatory, explained to the local newspaper when a new vehicle that ran on liquid propane was delivered to the scientific facility for use by the staff in the hopes of solving the "spiking" problem.[14] Fortunately, the propane car was successful in significantly reducing the emissions of particulate matter, carbon monoxide, and hydrocarbons.

But what had Dr. Pueschel meant when he said "to keep us alive?" Was he talking about the observatory's need to keep the air they sampled clean in order to justify and "keep alive" continued funding of study atop Mauna Loa? Or was he talking about the need to maintain the purity of our atmospheric air, to "protect our environment" in order to keep humanity alive? The newspaper reporter assumed that it was the first interpretation. Now that we know more, it isn't so obvious.

The monitoring of air samples at Mauna Loa was able to reveal a trend that has continued predictably over time. Since 1955, when monitoring began, to the present, the level of atmospheric carbon dioxide has risen somewhere between 0.5 and 2.88 parts per million per year. Each year, the blip in the graph that indicates the change of the seasons takes the line to yet a higher level, irrespective of whether there was a car driving up the side of the mountain or not.[15] Global air contains a little more carbon dioxide every year, and the 1998 reading was the biggest increase in a single year ever recorded.

The propane car at the observatory was eventually abandoned when an increase in traffic of regular cars up the mountain made it pointless to attempt to maintain the purity of the air to the same high standard. Yet despite the fact that the liquid propane needed to run the car wasn't readily available in 1972 (and it cost more than gasoline) and that the expense involved in converting the engine of a single car to propane was about $600, Dr. Pueschel felt that conversion was worth the additional expense, even for the average car owner. Dr. Pueschel knew something that virtually no one else knew: what a single automobile could do to air quality. He personally had witnessed the impact a lone car had made on the monitoring equipment. "Someday we will have to pay, and it won't be cheap," he predicted. Dr. Pueschel had also seen the relentless upward climb of the needle measuring carbon dioxide in the global atmosphere during the previous two decades. Could his warning have been any more clear? "We take for granted our air is free, but someday we just won't have it anymore."[16] The wisdom of hindsight may illuminate what he really meant when he told the newspaper why they had bought the propane car.

Gradually, incrementally, we are changing Earth's atmosphere. But are we slowly altering our atmosphere away from something that supports human life toward something deadly like the atmosphere of Mars? Such an atmosphere would have been very familiar to Joseph Black, who isolated the very first atmospheric gas. Unitarian minister Joseph Priestley would have recognized the atmosphere of Mars as well. So would coal miners from the early part of the 20th century and the canary that lay gasping at the bottom of the cage, for the atmosphere of Mars is made of fixed air. The atmosphere of Mars is made of blackdamp. The atmosphere of Mars is made of carbonic acid gas. The atmosphere of Mars is made of a substance that has over time had many names reflecting the toxic side of its nature. While today we call all of them "carbon dioxide" (which we think of as a benign product of our own bodies and the harmless bubbles in soda pop), this substance has clearly not always been viewed as a harmless gas. Nor should it be in the future, for it is time once again to inform our opinions about this substance and recognize its invisible, dark side. As long as a stylus attached to the monitoring equipment in some lonely station on the top of an inactive volcano in Hawaii continues to etch a line ratcheting upward—showing the increased amounts of carbon dioxide that, year after year, flood our atmosphere, threatening us—then we too must think of it very differently.

It isn't a matter of speculation. It is a matter of hard, cold scientific fact supported by numerous studies conducted by many respected scientists.[17–20] In the overwhelming majority they agree: Earth's atmosphere has far too much of what we now must think of as carbon die-oxide. It is warming our planet to the point where life, human life, is endangered. We are going to have to do something decisive and effective about this killer. No matter how successful or enlightened we think ourselves to be, we are not exempt from the need to act—in the same way that we are not exempt from the need to breathe.

CHAPTER 3

A Whole New Universe

I t took weeks for the gift to arrive by special shipment, and, had Dr. Robert Koch's wife known how much time he would spend with his eye glued to the strange device, she might have used the pfennigs she had so carefully saved to purchase something else. Never let it be said, however, that Emmy Adolphine Josephine Koch didn't make the right gift to her husband and childhood sweetheart for his 28th birthday. The coveted Hartnack microscope he carefully unwrapped on December 11, 1871 arrived not a moment too soon, for on the following morning, as the astrologers predicted, there was a solar eclipse at dawn, and as every astrologer can tell you, a solar eclipse indicates a new start in life. No doubt the morning light—when the sun returned—found physician Robert Koch looking at some tiny, twitching creatures through the lens of his brand-new microscope.

Before long, the little rented house the couple shared with their four-year-old daughter, Gertrude, included a makeshift laboratory. Although it had only four rooms plus a small study piled high with books and papers, it was light and sunny, and the large dining room had a wide bay window with an excellent view of Wollstein (which was then in Germany) in three directions.

Gertrude was to describe her father's laboratory this way. "My mother divided the large room in two with a brown curtain suspended on a long pole. The smaller half in the rear was fitted up as a workroom. This was my father's first laboratory, in which he could work more or less undisturbed. At the window was his equipment for making microphotographs and he hired a carpenter to build the necessary dark room. It stood there like a big closet with a black curtain hanging in front of it. Next to the dark room was an incubator. On the opposite side of the room was a small table laden with photographic apparatus, a microscope, and covered glass containers. In each of these sat a white mouse, ready to be used for some experiment."[1]

Emmy, who frequently acted as Robert's laboratory assistant, might be a candidate for sainthood someday, for she shared her home not only with her husband's laboratory, but also with its contents: acrid chemical solutions, numerous white mice, and most probably tens of millions of other tiny lifeforms, invisible except under the lens of the new microscope. Most alarmingly, these included the wiggling rods and threads of the deadly Anthrax bacillus.

While the notion of sharing a residence with the source of Anthrax may give today's reader the cold sweats, in the Germany of the 1870s, as with the rest of the world, contagion—particularly by something that was invisible to the naked eye—was not only an unproved concept, it simply wasn't believable. The notion that you could catch a fatal illness from some tiny, virtually invisible animal must have seemed like hallucinatory fiction, a nightmarish hoax designed to dupe the gullible. In retrospect, we might understand her husband's noble intentions, but Emmy might have had trouble convincing her friends and family that what Robert was doing with his microscope night after night was anything but foolish.

While Robert was busy proving that these tiny animals transferred disease (and he was the very first to do so), Emmy, who ran the household as well as assisting her husband, was busy making ends meet. Until young Dr. Koch was able to convincingly demonstrate that these tiny creatures (microbes) were the cause of disease—which took many years—he funded his own research with his private practice. Hunting for microbes was something he did on his own time, for the love of it, and it was nine years before his pioneering work was rewarded with an appointment by the German government as an associate of the Imperial Health Office in Berlin. With his appointment, for the first time he had a well-equipped laboratory and two assistants.[2]

Over the course of his remarkable career Robert Koch was to travel all over the world in pursuit of the microbes which caused humans and animals to contract diseases such as cholera and tuberculosis. In many instances he was also able to devise methods of prevention or cure. Time and again he headed off from Berlin: to Alexandria, at the request of the Egyptian government to help stem a cholera epidemic; to Calcutta, India, where he first isolated the microbe that caused cholera and then identified polluted water as the source; to Africa, to study the cause of a disease in cattle and on a second occasion to find a means of fighting sleeping sickness in humans. These were just a few of his accomplishments. In 1905, the day after his 62nd birthday, Robert Koch was awarded the Nobel Peace Prize for his immense contribution to humanity.

While Koch was working at the prestigious Imperial Health Office, which was funded by the Kaiser, he discovered one of the nuisances associated with the money he received: auditors. As could be expected, his laboratory was audited from time to time. On one such occasion the auditor general had every reason to be pleased with the way the Kaiser's money was spent. People were busy doing good work. Some were dissecting dead animals, others feeding and caring for the live ones, yet others were administering medicines, and a few were peering into microscopes and Petri dishes. Only one man, hands behind his back, was idly pacing back and forth in the corridor. The auditor wondered about it and asked his escort, "What in God's will is this man doing? Shouldn't he be busily working like the rest of you?" "Dr. Koch, sir?" the escort replied. "He's doing the most important job. He's thinking!"[3]

While thinking may be the most important job of a scientist (despite what auditors may figure), any number of scientists would agree that thinking is sometimes the bane of their existence, leading them on pursuits that are troublesome. It's like a virus. Once you've been exposed to certain thoughts the infection must run its course. While for Dr. Koch, thinking initiated nine years of a self-funded scientific quest, and ultimately led to a clear and grand victory, sometimes the battles are longer, the victories less certain and the quest even more heroic because of the huge political and cultural violations implied. (Galileo comes to mind.) Nevertheless, thinking is what starts all the trouble, and once the inquiry has been launched, some are compelled to follow the path wherever it takes them. Take Vincent DiPietro. Fortunately for Vince, as all his friends call him, by the late 20th century physical torture had pretty much been abandoned as a method for reining in scientists with heretical notions.

Maybe he caught the thought virus from his co-worker at Goddard Space Flight Center. On a dreary December morning in 1979, Vince, an Italian American engineer specializing in satellite transmissions for the NASA contractor, was at the bench examining some newly constructed circuit boards, subjecting them to routine testing, when his co-worker, Ben, dropped by to exchange a bit of office gossip. However, the animated man who showed up that morning was not Vince's normally relaxed friend. Ben was excited about something. Very excited.

Ben began to spill a very strange story with such fervor that even Vince (who would normally shrug off such a tale) had to pay attention. He listened quietly, if dubiously, while Ben, gesturing for emphasis and drawing a little picture on a scrap of paper, told Vince about what he'd seen. "It's a face, a giant face!" he claimed. "Right on the surface of Mars. No kidding!" Vince, who wanted to believe the story, also found it extremely unlikely. Ben clearly got the message. "Go check it out yourself if you don't believe me," he suggested. Vince realized, as his own inner circuits registered a moment of icy fear, that any such "checking out" would generate a lot of ridicule if someone found out about it. While he listened to his friend, he did nothing more.

But the fates had a different plan: while waiting for his wife to try on clothing in a shopping mall a few weeks later, Vince was thumbing through magazines in the bookstore. Then he casually picked one up that offered to show "pictures from Mars" and, turning to the aforementioned article, his attention was immediately drawn to a single compelling image. There was a "face." Transfixed he found himself staring at a black and white image of a serene human-like "face" against the background of the Martian surface. "Is this what Ben was talking about?" he wondered.

While the background of the image looked authentic, the "face," Vince reasoned, could easily have been pasted onto a picture of the Martian plains. An interesting fabrication, he concluded. He looked over the other pictures the magazine had published. Although they seemed to be fairly standard images of the type returned from Mars by the Viking mission, the captions included a number of inaccuracies he easily recognized. Based on what he was looking at and already knew, the article was either the effort of an overly zealous and under-informed writer or a pure hoax, and he couldn't tell which. He returned the magazine to the rack and wandered off to find his wife, not realizing that the memetic,[4] or "thought virus," Ben had infected him with was no longer merely incubating. It had now taken hold. Vince was thinking and they were the thoughts of a very curious space-age technologist.

In a sense, the modern shopping center where Vince was looking for his wife is an embodiment of space-age technology. Most of the "malls" built in the last 25 years in Asia, Europe, North America and elsewhere, look like the complexes in science fiction worlds. Inside large, fanciful constructions—many with sheet-like glass walls and skylights, marble floors and stainless steel handrails—passengers in crystal-like elevator cars push buttons and levitate through open, tree-filled atriums, or glide gracefully up and down the slanting ramps of whispering escalators.

The air inside these shopping centers is usually "conditioned" to a perfect temperature. Even in the bleakest winter, you can walk comfortably in dry, windless warmth without a coat; in hot weather, you can stroll in cool, perfect comfort. Space-age wonders abound, such as doors that slide open when our presence is detected, or taps that magically dispense water at the perfect temperature when we thrust our hands beneath them.

The variety of merchandise in the various stores would be mind-boggling to someone who had lived a hundred years ago. The colors, forms and functions of the goods are a dazzling testimony to the progress of the last century. Even the materials would be unfamiliar: silicone chips, magnetic media, high-tech plastics, and exotic synthetic fabrics.

In fact, nearly everything about the modern shopping center would be unimaginably futuristic to a visitor from a hundred years ago. You would have to explain what everything was: the cars in the parking lots, the bar code scanner, the security system, your credit card, even the hand dryer. At some point, this visitor from the past might look at you in awe and ask, "What made all of these wonders possible? What is the power behind all of this?"

After trying to explain by flipping a few light switches and pointing to a number of electrical outlets, your guest indicates that this isn't exactly what he wants to know. "Where is your dynamo? Your power plant?" he asks patiently. Finally you understand the question. So, you lead your guest outside the building to the parking lot, then back behind the mall, beyond the trash bins, to a place that almost no one ever sees. (This is where you really have to use your imagination because this place isn't usually situated so close.) There, behind a chain-link fence, is a dark hole in the ground next to a big brick building with a very tall chimney. A variety of cables, a pair of rails, and a conveyor belt disappear into this black void. You point to the hole and explain that the power behind all of this amazing technology comes from a black, carbonaceous substance found deep in the ground.

Your visitor from a hundred years ago looks at you incredulously. "So you power your miracles with coal," he stammers. "You still use coal?"

You nod affirmatively. "See the coal moving into the power plant?" You point to the brick building with the very tall chimney that sits next to the mine. A conveyor belt loaded with coal runs directly into the side of the building.

"You have all this amazing technology and you're still driving steam turbines with coal?" asks the guest, trying to hide his amusement.

"You have a problem with that?" you respond a bit defensively. "Hey, we use coal to generate this really cool stuff called 'electricity.' See, there're these wires ..." You suddenly notice, with a flash of consternation, that your guest is laughing at you. Here is a person who didn't know how to push the button on the hand dryer just a minute ago, yet who seems to be mocking you now. How did he get so smart all of a sudden when half the people you went to high school with don't know about the coal-powered process that generates electricity? The mystery is solved when you realize that the process of generating electricity by burning coal would have been very familiar to your guest. After all, humans have been generating electricity in almost the same way for over a hundred years.

Electric lighting was first introduced in the 1880s. At the beginning theaters, stores, companies and wealthy citizens who wished to electrify their homes or establishments had to install their own electrical generators. However, as the popularity of electric lighting took off, so did the building of coal-burning power plants to supply the necessary electrical power. Soon cities were criss-crossed with electrical lines and the air was filled with fumes.[5]

Here is a description of a modern power station. "At United Illuminating Company's harborside Unit 3 in Bridgeport, Connecticut, built in 1969, Kentucky coal rides in on a conveyor belt to five huge blue silos that can each hold 700 tons. When pulverized to the consistency of talcum powder and burned in UI's huge boilers at 2,500 degrees, the product is steam, which turns a humming General Electric turbine at 3,600 revolutions per minute, generating 400 megawatts of electricity for county consumers."[6]

Ironically, that whole space-age shopping center that appears to be so technologically sophisticated is really driven by a hidden, antiquated, 100-year-old technology: a steam-driven turbine. To an embarrassingly high degree, all of our "space-age" technology is built upon a foundation of much older, dirtier energy technologies, and that very foundation is weak and threatens to fall, perhaps taking a lot of space-age technology with it. If we had updated our "back end" energy sources at the same rate as we developed the "front end" technologies that plug into the wall juice, we would be producing clean, inexpensive and, most importantly, nearly carbon-free energy. However, we've neglected to be as innovative in the domain of energy technology as we have in everything on the "front end." Even when we have developed a relatively modern technology, such as nuclear-fission-generated power, we haven't gone nearly far enough to make the technology both clean and safe. And more to the point, we even stopped implementing our clean and safe innovations.[7] So, the gleaming futuristic-looking facility we admire so much is a mere facade which hides the dark truth of a giant, antiquated, 19th-century fossil-fuel-burning stoker in the backyard.

In the case of most shopping centers, a coal mine is nowhere in sight. Neither is the coal-fired power plant. Of course, some of them aren't even powered by coal for that matter, although the vast majority are. According to the The World Resources Institute, electricity in the United States in 1995 was produced as follows: 70.11 percent was produced by coal or gas; 9.22 percent by hydroelectricity (dams); 20.13 percent by nuclear fission; and 0.54 percent from naturally occurring geothermal sources.[8]

The process of generating electricity is invisible to most Westerners. In fact, both the coal mine and the power plant could be so far away that you could live your whole life in a town and never see them. However, they are nearly always there. They may be totally invisible to you, but you need to know that somewhere out back, metaphorically speaking "behind the trash bins," is the coal mine and the power plant that are actually running the place. All that shows up most of the time is the bill. The rest, including the 6,305 million tons of carbon dioxide that were released into the atmosphere worldwide during a single year, is invisible.[9]

Carbon dioxide may be invisible, but that only limits its visual impact, not its physical one. Everything goes somewhere. Any "vanishing," on the physical plane of reality, is an illusion. Matter may change states—frequently to the point where it is no longer recognizable as its previous form—but it doesn't just disappear. It persists in some form. On the one hand, this may be obvious. On the other, it isn't obvious at all. "Where does it go and what does it do?" are critical mysteries that must be solved before any chemical or microbe is unleashed upon a planet.

Many of our communications about everyday life perpetuate our illusions about invisibility. When matter makes a state change from something visible to something invisible, we actually talk about it as if it had "vanished." When we blow out a candle, we say that the candle is shorter, not that the room is full of smoke. When the cube melts, the ice is "gone." We throw something "away." The bottle is empty, rather than filled with an invisible substance. When we drain the bath tub, we certainly don't say we are displacing the water with air. The firewood we "burn" doesn't change to smoke and heat. It seems to disappear.

While there is nothing particularly unusual about this state of affairs, nothing sinister in an empty bottle, it may be useful to notice how pervasive these sorts of "small illusions" are. Hundreds of these tiny, harmless misrepresentations are part of our everyday thinking.

When we speak of the invisible as having vanished we are perhaps using a workable "consensus reality" that allows us to communicate with other humans in the way in which we are all accustomed. For example, in our culture it would seem very strange to ask for a glass filled only with air, even if that's technically what we wanted.

But there is a dark side to this consensus reality: this form of self-deception also allows us to ignore invisible substances which are very harmful. By failing to "see" these substances, we allow them to surround us, sicken us and even kill us without ever acknowledging their reality. We can breathe, drink, and ingest them. We apply them to our skin, our clothing, our appliances, our floors, and in various ways allow invisible toxins to come into contact with our bodies, our children, our homes. We allow those who benefit from releasing invisible toxins to continue to do so. Frequently their acts are unknown and unchallenged, like an enemy fleet that has flown in under the radar. If something is invisible, it is frequently beneath our threshold of awareness. And, if we are unable to perceive something, in our simple innocence (dare we say ignorance, which interestingly enough shares the same root as the word ignore), we act as if it doesn't exist. Thus, both our missing perceptions and our minor, culturally-inherited self-deceptions allow us to harm and be harmed, both directly, by poisoning our bodies, and indirectly, by poisoning our environment.

Consumer advocate Ralph Nader put it this way nearly a quarter of a century ago:

> Basically, people come equipped with the ability to detect certain dangers—smells, sight, hearing, taste, thresholds of pain. Man is geared up to avoid fire. Fire burns. Man says "Ouch," then runs away or puts it out. But now, human beings are producing fires that burn over a long period of time; they don't burn immediately and we're not set up for it. We have to develop systems—legal, medical, democratic—to detect these dangers before it's too late for a lot of people.[10]

While the illusion of invisibility is true of many things, it is particularly true of air. We can perceive wind, and temperature and clouds, but the air itself is "nothing." It isn't even merely transparent, like a clear glass. Air is a void in human perception. It is an empty backdrop to everything that is, keeping things separate from each other maybe, but air itself isn't normally thought of as a "something"—it's "nothing but air." At best, air becomes "only air." But most of the time, air isn't up to even that level of "is-ness." Most of the time, air "isn't."

Imagine that you can perceive the air and notice where those thoughts take you. Be conscious of what fills your lungs, surrounds your body, stands between you and any other thing. Can you feel its presence connected to your flesh? Are you aware of how essential it is to your survival? We are like fish in water. We swim in air.

One of the most important recent scientific discoveries was all the more remarkable because it involved searching for one invisible substance (a man-made chemical) inside of another invisible substance (the atmosphere). Even stranger, when the scientists made their remarkable discovery, they found that the invisible substance they were searching for had made an invisible, but deadly, hole in the invisible substance they found it in.

Mario J. Molina, who had recently finished his Ph.D. in chemistry, joined Professor F. Sherwood Rowland, a specialist in radiochemistry, at the University of California in Irvine in 1973. Together they sought a group of invisible chemicals commonly known as CFCs, or chlorofluorocarbons, a form of chlorine gas. CFCs were widely used in industry as a propellant in a vast variety of aerosol canisters; everything from spray deodorant to spray paints. Somewhere in the vicinty of a billion cans were produced every year.[11] CFCs were popular for a very wide variety of uses because they were chemically "stable" which meant that they didn't deteriorate. They also didn't seem to react chemically with anything else, so they would not chemically change the substances they were propelling. However, it was this non-reactiveness that was part of the problem. Whereas some industrial chemicals begin to "break down" into less harmful substances when exposed to the abundant chemicals of our environment, those of the air and water, CFCs did not break down in the presence of ordinary chemicals.

While this fact alone was interesting, Molina and Rowland were also curious about a scientific study which seemed to support the notion that CFCs were virtually indestructible. A recent measurement of atmospheric gases had indicated that all the CFCs that had ever been released were still in the air.

Rowland, however, suspecting that *something* must eventually happen to CFCs, decided that this was a worthy target of investigation—a scientific mystery. What Rowland and Molina found was that over many years the CFCs slowly migrated upward to the high atmosphere where they then initiated a catalytic chain reaction that caused the destruction of over 100,000 ozone molecules for every molecule of CFC. The ozone layer, which shields the Earth from the ravages of the Sun's ultraviolet radiation, was being destroyed wholesale.

While it took years of effort to enroll governments into regulating CFCs, since industries chafed at the economic losses, an interesting alliance between scientists and consumers won the day. The scientists (in Europe, notably Paul Crutzen), who knew about the dangers of CFCs, spoke out loudly enough to be heard and with such force of intention, that even in the face of corporate displeasure, we all became aware of what was at stake. Science became a "force" to be reckoned with. Not since Rachel Carson's amazing book, *Silent Spring*, was released in 1962, alerting the world to the dangers of the pesticide DDT, had the public been so galvanized by an environmental issue. Consumers, who took direct action by refusing to purchase products containing the offending substance long before companies were required to remove them, made it clear they would not tolerate the further destruction of the ozone layer. Together, scientists and consumers were able to move the issue to the forefront of the media. The combined effort was effective. Aerosol canisters were literally pulled en masse from the shelves because, at least then, we acted as if the health of humans and the safety of our planet were paramount. Quick action saved many lives. It may have saved yours.

At the same time as this issue of vast atmospheric damage was being addressed by one group of scientists, another group was looking at the oldest, smallest atmosphere we had ever seen.

Like archeologists trying to read unfamiliar hieroglyphics, a group of scientists with the U.S. Antarctic Meteorite Program, a joint venture between the National Science Foundation, NASA and the Smithsonian Institution, were trying to read the return address on a meteorite they'd discovered, which had been lying on the ice and snow of Antarctica for more than 10,000 years. They named the meteorite, the first one found at Elephant Moraine in Antartica, in 1979, EETA79001. Immediately it was recognized as a rare specimen. Unlike most ordinary meteorites, which are chondrites, and consist of a shock-welded mass of rock fragments and space debris apparently left over from the formation of the solar system, EETA79001 was an achondrite. Not a conglomerate at all, it was volcanic lava. More remarkably it was actually composed from two lava flows, one on top of the other, with a vein of calcite in the plane that separated the two. Its outside was covered with melt glass which had formed when it was torn from a larger celestial body. In the glass were numerous small bubbles of trapped gas (these were discovered in 1983) that might under analysis reveal the meteorite's origin.[12]

Even more intriguing, this unique rock contained a vital clue which, if unlocked, could solve a mystery about the whole category of strange meteorites to which it belonged. They were known as the SNC, after the locations where three prominent representatives of the group—Shergotty, Nakhla, and Chassigny—were found. Shergotty fell in India and was an iron-rich basalt, like lava from Hawaii. Nakhla was a calcium-rich lava, and Chassigny was composed of olivine, a green jewel-like volcanic rock. All of the SNCs were remarkably young by meteoritic standards. The SNCs age of crystallization could be determined by measuring their radioisotopes. Unlike most other achondrites or chondrites, which had ages of nearly 4.5 billion years, the ages of the SNCs ranged from 1.3 billion years down to a geologically "young" 160 million years. They had to have come from some body in the solar system that had geologically recent active volcanism. It was speculated that parent body might be Mars.

From the Mariner and Viking pictures, it was known that Mars had young volcanoes. However, no one could believe that an impact on a planet as big as Mars could propel rocks clear off its surface and into space. Even more difficult to comprehend was the fact that the rocks would then have had to orbit the Sun for hundreds of millions if not billions of miles before landing on Earth. It seemed reasonable to suppose that any explosion on Mars powerful enough to have that effect would have destroyed the rock rather than boosting it into solar orbit. So, since it seemed impossible that the meteorites had come from Mars, the assumption was that they must have come from some asteroid in the main asteroid belt.

In May 1981, a piece of the Moon was found in Antarctica and shattered the reasoning that finding Martian meteorites on Earth was impossible. It was a rock almost identical to those brought back by the Apollo astronauts, but it had reached our planet without the aid of a rocket. It had apparently been blasted off the Moon by a meteor impact. Suddenly the idea of big rocks from space (comets or asteroids) hitting planets and blasting other rocks back into space without melting or pulverizing them in the process seemed feasible. It had already happened on the Moon.

The idea of the SNCs being from Mars suddenly gained new life. Finally, when the gas captured in the tiny bubbles in the glass of EETA79001 was subjected to analysis and found to have an isotopic profile virtually identical to the atmospheric gases that were sampled by the Viking mission on Mars, we knew that this small visitor from outer space was really a piece of Mars.

If ever there was clear evidence that "air," although invisible, contains a great deal, this was it. Inside a tiny gas bubble aboard a meteor, there was enough substance to identify both the meteor and its trapped air as being from another planet over 100 million miles away. If an iota of air can say this much, consider what the contents of someone's lungs might have to say.

By the time Christy Booker was in secondary school, she was taking a new medication for her asthma, one that she would take on a maintenance basis for many years: Theophylline. While this medication helped to control her bronchial spasms, and it could be taken at home, it had a highly stimulating effect that was similar to the adrenaline shots she used to have from the doctor when she was in a major respiratory crisis. In her own words, this medication would make her "crawl the walls."

In some ways the side effects of Theophylline, as it was used at that time, were worse than the adrenaline. With the adrenaline, once she was breathing more easily, she was given another medication to alleviate some of the side effects. With Theophylline the side effects were unrelenting. She was given the medication every 12–24 hours in gradually increasing increments over a period of two weeks. Every two weeks a sample of Christy's blood was drawn at her doctor's to monitor the levels of the drug in her system. Any adjustments in the medication were also made in slow, gradual increments. During all of this, Christy's neurophysiology was going completely haywire.

As much as she wanted to, sitting still at a desk in school was nearly impossible for Christy. Unable to concentrate, unable to complete assignments, her academic progress suffered.

Ironically, while she was thrumming with medication that made her want to run, running, and, in fact, any rigorous physical activity, was forbidden her. Like many asthmatic children, breaks between classes and gym were spent sitting on the sidelines. Between the physical and consequent social isolation–imposed by both the asthma, and the effects of the drugs—Christy's social development was slowed and she was labeled as a child with a behavior problem, lacking in self-control. Had Christy not been so medicated, school-life might have been another kind of experience for her. However, mood-altering, life-altering medication on a daily basis was part of all of her early years and she will never know how it might have been otherwise. But then, none of us knows what we don't know.

There is a kind of "hole" or missing ability in human development which makes it easy for us to remain oblivious to clear and present dangers, and to act as if they don't exist at all. For humans, there are two broad categories of "things." There are the tangible, physical things like tables and trees and eyeglasses. Then there are the intangible, conceptual, abstract and non-physical things like how much you love your mother, a decision about your future, and the recognition that slavery is bad. We live in a world in which we regularly deal with both kinds of "things."

It was Jean Piaget (1896–1980), a biologist and naturalist, who laid the ground work for how we learn to relate to things both tangible and abstract. His theories on the stages of cognitive development in early childhood illuminated our understanding of how people eventually evolve to adult thinking patterns. Although he was not trained as a psychologist (most of his early work was with mollusks), Piaget noticed that as children grew, they not only knew more but they actually thought differently about what they knew. Over time, he was able to outline the sequential stages typically followed in the development of children's thinking.

Piaget defined the result of this progressive refinement, which automatically occurs as children experience more of the world around them, as a "schema," or inner "map," of the world, which alters as a child fills in different levels of understanding through experience. The schema of a young child is mostly undeveloped. So, given exactly the same information, a person using an adult-level schema will relate to it far differently than will a child.

·One of the chief characteristics of our very earliest stages of mental development is the inability to mentally sustain "object permanence." This is basically an elaborate way of saying "out of sight, out of mind." In order to keep the memory of something in mind, you have to have a mental symbol or image of it, something a very young child doesn't yet have as part of its schema. This is why, when you show a small infant an object that has attracted her attention and then block it from view, she will act as if the object has simply vanished. She won't make any attempt to find it, even if it is a favorite toy. So parents regularly entertain (or upset) a small infant by making an object "appear" and "vanish" by simply covering it up.

Playing peek-a-boo becomes even more fun when a child begins to understand that the object you are hiding hasn't really disappeared and begins to look for it. Piaget himself noted that when his infant son, Laurent, was 7 months and 13 days old, he moved his father's hand aside to get to a hidden matchbox, a clear advancement from earlier games where Laurent had never made such an attempt.[13]

At a later stage, a child will look for an object where it has been hidden most frequently, even if he's observed it being hidden elsewhere. If you hide a toy under a blue cloth repeatedly and then in plain view put the toy under a red cloth, the child at this stage will look for it under the blue one. Only at a later stage will the child look for a missing toy in the last place the toy was hidden.

It takes years of development for human beings to reach the point where we understand that things don't just disappear because we can't see them anymore. Yet, in some ways, this stage of "object permanence" never gets hardwired into our everyday reality. While we aren't quite as gullible as children watching a magician perform a sleight-of-hand card trick, nonetheless, when the garbage collectors empty the trash cans into their truck, or we flush the toilet, we generally relate to the process as if the trash and sewage just vanished, never to be seen or thought of again.

There is another aspect to object permanence, though. How do we relate to an object if it is invisible? Where does an invisible object go when you hide it? Does it stay where you put it? How can you tell? How do we relate to an object or substance outside of our normal range of perception? We relate to it as if it doesn't exist. Like a baby who lacks any knowledge of potential threat, if you were about to be hit by an unseen and, therefore, invisible meteor, you wouldn't even flinch.

The point is this: our perceptions are limited. We are literally blind to some the most serious threats to our well-being. That's one of the reasons we can tolerate tons and tons of harmful gases being emitted from cars, for example. If we could see the stuff, if it came shooting out of tailpipes like poisoned arrows, we'd grab the kids, run inside the house and call the cops. Instead, we can blithely drink water laced with chemicals and feed our children hormone-laden meat without even a minor sense of alarm. We sit there innocently, taking it in the chops, while we get cancer and asthma and kill our planet with global warming. We don't even make the connections to the causes. Like a high-frequency sound that dogs hear and we don't, our limited ability to sense leaves us ignorant, vulnerable and easy to exploit.

According to Piaget, the last stage of cognitive development in humans involves developing the ability to think and reason in the abstract. This is the point at which we begin to be able to manipulate concepts and ideas—things that are not part of physical reality. By developing the ability to think in abstractions, our schema grows to include non-physical reality. Through our ability to visualize and imagine, we are able to generate mental pictures that expand our schema. We learn to "see" with our minds what our eyes can't show us.

The only way that we learn to avoid dangerous, invisible things in our environment is to learn to think about them abstractly, and to somehow act as if they were visible. A contagious disease is a good example. It is impossible to see a virus or bacteria with your eyes. If you were to relate only to what you could see, you would be far more likely to drink water that gave you cholera or to have unprotected sex with an AIDS-infected partner. We have therefore learned to act as if there really is something physical present, something that could harm us, even if we can't see it. We've learned to be careful about washing our hands, our bodies and clothing. We've learned to avoid drinking water that may make us ill. We've learned to use a condom as a barrier to invisible viruses. We've also learned to protect other people by covering our mouths when we cough or using a tissue when we sneeze. We vaccinate our children, bandage their cuts and teach them personal hygiene.

While this all seems so obvious, so much a part of everyday life, that's only because we've already learned to think this way. Ask any mother who has taken a child through all of the steps of learning personal hygiene—this is not a simple task. Until very recently in human history, all of the habits that we consider essential to good health would have been viewed as absolutely unnecessary—and maybe even counter-productive—since they are all strategies designed to protect us from threats which are totally invisible to the naked eye.

We must learn to treat invisible environmental contaminants with exactly the same alarm and caution that we show toward *E. coli*. We must learn to treat carbon dioxide as a waste product. We must be as rigorous in teaching our children to avoid things tainted with harmful chemicals as we are in teaching them about germs. We must encourage others to do the same. We need to treat those who would contaminate us with harmful chemicals with the same contempt that we would feel for someone who would intentionally spread AIDS. We can transform our cultures quickly if we are intentional. There is no excuse not to be. We can no longer tolerate this form of ignorance in ourselves or others. The next step in human evolution requires this new standard for survival.

For example, it wasn't that long ago that a physician would leave an open cadaver at an autopsy and go to deliver a baby without washing his hands. What possible justification could there be today for a physician operating with unwashed hands—now that we know the truth?

While Earth had captured a piece of Mars, it seemed that Mars had captured something on Earth—Vince DiPietro's attention. His life-long quest for truth was taking a giant step forward, even if it only looked like he was taking a brisk lunch-hour walk on a wet winter day in Maryland to building number 26 of the National Space Science Data Center (NSSDC). Although it had been several days since his visit to the bookstore, he hadn't been able to shake off the effect of what he'd seen in the magazine. He decided to do something about it. He was going to find out, to his own satisfaction, by examining the archives of the Viking Mission, if there was anything to this "face" on Mars. However, after seeing the thousands of files and rolls of film there, he realized he would have to return to do an extensive search. The next morning he took a "vacation day" and spent it poring over the Viking archives in the dim basement of the NSSDC.

It wasn't long before his efforts paid off. While sifting through materials in the three-year-old archive he found an official press release photo that looked exactly like the one in the magazine. So perhaps the image wasn't a fake after all. Here it was in the official archives, neatly labeled 76H593/17384 with a single word of annotation: HEAD. With the help of the very cooperative staff, he was soon sitting at a film viewer examining a frame of film, number 35A72, that revealed the image in much greater detail. Even on closer examination, the facial features were clearly visible. While its presence was in one sense amazing, inviting further investigation, in another sense Vince felt relieved that his job was over. NASA knew about the feature and would, undoubtedly handle the investigation with their typical professionalism. On the way out, he stopped to order a photographic enlargement, adding to his sense of satisfaction at having completed his mission.

Soon after the arrival of the enlargement he'd ordered, DiPietro reached a stunning conclusion about this Martian feature that seemed to be staring back at him so clearly. No one had ever studied it. From the day the digital data showing this unusual Martian feature had arrived on Earth, no one at NASA had ever investigated it, beyond the cursory glance it was initially given by Tobias Owen with a magnifying glass while looking for a landing place for the Viking lander. Three years earlier, on the very day of its discovery, Dr. Gerald Soffen had dismissed the "face" as a "quirk of light" which had vanished in a second image taken a few hours later. Then the matter had been tossed aside.

Vince found a sympathetic listener in a fellow employee at the Goddard

Space Flight Center, Greg Molenaar, a Swedish-American computer scientist, who like Vince was an avid follower of the space endeavor. As they examined and discussed the image together, their excitement and sense of purpose grew. It dawned on them that since no one else had done anything to verify the image, it was up to them. In that moment a wonderful mission was born, instantly recapturing for them the excitement and sense of grand purpose of the Apollo program when humankind pursued the quest to land upon the surface of the Moon. Since no one else had bothered to do the science, Vince and Greg were going to explore the surface of Mars—on their own.

Greg immediately suggested a method of improving the image quality using a computer process. In a sense, images from space aren't "photographs" (exposed and processed film) at all, rather, they are digital scans stored as electronic files of pixel values which are recorded by cameras and then transmitted back to Earth and stored as digital information. Greg and Vince requested the original digital tapes from the Jet Propulsion Laboratory (JPL) in California. They intended to obtain the tape and study the image pixel by pixel. Yet even as Vince and Greg prepared, lining up film recorders and computers, a nameless fear began to grow, a fear that in requesting the tape and pursuing this inquiry, they were doing something that was forbidden. Yet there was no turning back.

Like all good investigators, Vince DiPietro and Greg Molenaar went into information-gathering mode when the tape with the Viking data finally arrived. The questions about the image they were investigating loomed large, but all speculation was at a standstill. They were putting their energy into finding answers and it was going to take a tremendous amount of work.

They began with NASA equipment at Goddard Space Flight Center. They asked, and were granted permission, to use the equipment—a computer nearing the end of its useful life, and some specialized image processing gear. Both amused and curious, a supervisor insisted on giving the permission in writing. "It might come in useful someday," he told them with a wizened smile. Without really considering the implications, Vince and Greg were launching the first serious scientific SETI (Search for Extra-Terrestrial Intelligence) investigation on what was potentially an alien artifact—and they were doing it in the heart of the aerospace industry, with permission.

Crammed into a tiny workspace with the esthetic appeal of a service elevator, Vince and Greg cobbled together the various pieces of electronic equipment to devise a serviceable, if dated, method of reading, processing and recording digital images. The newly-arrived tape could then be read on a tape drive and stored on the computer, which also drove a connected film recorder. The visual display was basically a cathode ray tube where the eight-bit pixels produced an on-screen image in 256 shades of gray. In 1979, state-of-the-art computers were slow and cumbersome by today's standards, and Vince and Greg's was even more so. Nevertheless, it worked, and before long they were processing the image and trying to find the correct film exposure settings by trial and error.

The first image was extremely gray with poor resolution and many "salt and pepper" transmission errors which were the result of interference in sending the data to Earth from Mars via satellite. After dozens of experiments, they were able to increase the contrast and correct for some of the transmission errors. However, resolution problems caused by pixelization and aliasing—what most computer users know as "jaggies"—continued to be a problem.

Undaunted, Molenaar and DiPietro developed a whole new technology for image processing which they called Starburst Pixel Interleaving Technique (SPIT) because of the way the transmission errors were accented like rays of light. The acronym "SPIT" was also apt since the result of the process showed "the spitting image." This computer imaging processing technique is actually a statistical analysis of data and resulted in a much higher resolution image.[14] However, as they stared at the enlarged image of the now familiar "face," they were overcome by a desire for further verification. They had to find another image.

Driven by the need for more information, Vince began to scour the Viking data looking for the "second image." After a month of systematic review in his spare time he finally did find a second high resolution image of the "face." It had been taken a full 35 days later and verified all the details seen in the first image. Taken from a slightly different sun angle, there were other details as well. Vince had no doubt that there really was a feature on the surface of Mars that looked like a "face." It had not disappeared. He ordered a tape of this second image from JPL immediately.

Although Vince was excited to find another image that confirmed the presence of the unusual feature, he was still worried. What had happened to the image that Dr. Soffen claimed showed that the "face" had vanished under different lighting conditions a few hours after the first one? In time he would learn that there was no such image in the archive because no second picture had been taken on the day the announcement had been made. Night had fallen on Cydonia after image number 35A72 was transmitted to Earth, so the "face" on Mars was covered by darkness for many hours after that. Of course, you could argue that in darkness everything vanishes—but nothing could refute the possibility that Vince and Greg were now looking at a second photo of something remarkable, something extraordinary, something that could forever change our view of the cosmos.

On the other hand, it didn't matter what the "face" on Mars might turn out to be. Because it had already achieved object permanence, the schema of life on Earth had opened to the schema of life in the cosmos and, together, our pair of Mars explorers plunged irrevocably into an unfathomed gulf.

CHAPTER 4

From Dust to Dust

Last weekend was the worst dust storm we ever had. We've been having quite a bit of blowing dirt every year since the drouth [drought] started, not only here, but all over the Great Plains. Many days this spring the air is just full of dirt coming, literally, for hundreds of miles. It sifts into everything. After we wash the dishes and put them away, so much dust sifts into the cupboards we must wash them again before the next meal. Clothes in the closets are covered with dust.

Last weekend no one was taking an automobile out for fear of ruining the motor. I rode [my horse] Roany to Frank's place to return a gear. To find my way I had to ride right beside the fence, scarcely able to see from one fence post to the next.

Newspapers say the deaths of many babies and old people are attributed to breathing in so much dirt.

from the diary of Ann Marie Low
Wednesday, April 25, 1934[1]

The dust bowl of the 1930s was the first large-scale environmental disaster ever experienced in the United States. It began with drought over a number of years, exacerbated by long-standing, environmentally damaging agricultural practices. It didn't end until it had lifted the topsoil from a vast region that ranged from Texas and the Oklahoma panhandle to the Dakotas, tearing the life out of the land across the central states of North American as far west as Colorado and east as far as the Allegheny mountain range in Ohio and West Virginia. Huge towering black clouds of dust would appear on the horizon and slowly engulf whole regions with the wind-blown silt, not just once, but over and over for month after month.

> Americans, always a profligate people, had disregarded the warnings of conservationists and ecologists. Contemptuous of the jeremiads of the "crackpots," heedlessly expending the natural riches that lay so easily at hand, Americans polluted the streams of the industrial East, cut down the evergreen forest of the North Central States, and wore out the soils of the cotton and tobacco South with generations of destructive single-crop farming.[2]

Although the disaster was self-inflicted, even as the winds blew the very substance of the Earth to the northeast, the farmers who still owned their land earnestly and optimistically tried to keep planting and replanting using the same bankrupted farming techniques—as if one single rainfall was all it was going to take to forestall a disaster that had been years in the making.

> Saturday Dad, Bud, and I planted an acre of potatoes. There was so much dirt in the air I couldn't see Bud only a few feet in front of me. Even the air in the house was just a haze . . .
>
> The newspapers report that on May 10 there was such a strong wind the experts in Chicago estimated 12,000,000 tons of Plains soil was dumped on that city. By the next day the sun was obscured in Washington, D.C., and ships 300 miles out at sea reported dust settling on their decks.
>
> from the diary of Ann Marie Low
Monday, May 21, 1934[3]

But the problem didn't go away. It lingered for year after year, like a slow not quite terminal disease. Swirling black blizzards brought silence and death to animals, plants and humans. It was as if the whole world were dying. Without water, without grass, weakened cattle, lungs full of choking dust, ate at the roots, slumped to the ground and died, their stomachs full of mud.[4]

70

The dirt quit blowing today, so I cleaned the house. What a mess! The same old business of scrubbing floors in all nine rooms, washing all the woodwork and windows, washing the bedding, curtains, and towels, taking all the rugs and sofa pillows out to beat the dust out of them, cleaning closets and cupboards, dusting all the books and furniture, washing the mirrors and every dish and cooking utensil. Cleaning up after dust storms has gone on year after year now. I'm getting awfully tired of it. The dust will probably blow again tomorrow . . .

July has gone, and still no rain. This is the worst summer yet. The fields are nothing but grasshoppers and dried-up Russian Thistle. The hills are burned to nothing but rocks and dry ground. The meadows have no grass except in former slough holes, and that has to be raked and stacked as soon as cut or it blows away in these hot winds. There is one dust storm after another. It is the most disheartening situation I have seen yet. Livestock and humans are really suffering. I don't know how we keep going . . .

from the diary of Ann Marie Low
Saturday, August 1, 1936[5]

It is important to be reminded of the back-breaking, heart-wrenching reality of this kind of disaster because this scenario is predictably one that is likely to be replayed over and over in the decades ahead as a consequence of global warming.

There is nothing in place to prepare us for this. Perhaps nothing can. We will witness the disasters wrought by decades of failed environmental policy and practice as surely as we were born. While it is less certain that it will occur in the middle of the United States, dust bowls will be happening somewhere, perhaps in many places around the planet at the same time. Whole regions that are now lush and green are likely to wither and die as a process of desertification or "making desert-like" steals across the tropics in an ever-widening band. Crops that support human and animal life will fail, repeatedly. But the climatic changes caused by global warming won't be limited to drought. Floods, hurricanes, and sudden violence in the form of winds, storms and tornadoes and wild temperature swings are predictable in the years ahead. It's all so sad, it makes you want to go somewhere else. Someplace like Mars.

W eeks passed and the second tape Vince DiPietro had ordered from JPL hadn't arrived. JPL was called. Yes, they had received the order, but the tape was not being released. "Why?" Vince and Greg wondered. Did they know about the second image of the "face" already? Vince and Greg Molenaar wrestled with their fear. Here they were, two "paycheck vulnerable" employees of a NASA contractor, pursuing what was both an avocation and a mounting obsession, with nothing backing them but the force of their curiosity. Now it appeared that the almighty JPL was opposing them.

Where curiosity began to fail in the face of fear, raw anger now took over. Armed with the realization that, by withholding the data, JPL was suggesting its possible importance, Vince and Greg pressed JPL relentlessly, demanding the publicly funded information that should have been theirs for the asking. They were charging the dragon's lair to plunder his treasures, and the dragon was aroused. But were they not brave knights?

Finally, Vince warned JPL that as taxpayers, he and Greg would ask their congressman to investigate the withholding of supposedly public data. It was JPL who actually served them, not the other way around. The response to this threat was immediate. The tape magically appeared. Nevertheless, the Mars boys felt it prudent to move the base of their investigation to a private contractor, LogE Tronics, in Springfield, Virginia.

The new site was 60 miles from Goddard and added another hour to their daily commute, and that was after their usual three-in-the-morning cut-off time. For Vince and Greg the problem of the "face" had gone from a curiosity to an obsession, demanding time and energy beyond all reason. Night after night until the early morning they labored, trying to bring enlarged, unsmeared, SPIT processed images of the object onto film at the correct exposure settings. All of this work was done by excruciating and exhausting trial and error.

Finally, one morning at three, they had both images of the object in comparable form: two portraits of a mile-wide object on the surface of a planet tens of millions of miles from Earth, taken on two different days under two different lighting conditions. They sat silently, contemplating the results. With their dawning recognition that, perhaps for the first time, two human beings could be looking upon a creation of non-human intelligence, came an almost paralyzing sense of dread, the feeling that they had obtained forbidden knowledge. Mars, it seemed, had been alive.

O n the day after Christmas 1983, I encountered something that forever transformed life for me. The day began innocently enough. I'd enjoyed a warm family Christmas with my wife, Faye, and three-year-old daughter, Elizabeth. Outside, in the beautiful northeast heights of Albuquerque, wisps of windblown snow swirled around the house. Warm and comfortable inside, I watched a teaser on TV for that evening's news magazine. It gave prominent mention to what they termed the "Face on Mars." I thought it was going to be a farcical piece that might also show some nice shots of the planet Mars, and I was eager for a diversion that day. The year that was drawing to a close had left me in an uneasy state of mind, and the year 1984 loomed a few days beyond, full of Orwellian uncertainty.

The year had not gone well for me. I was in my second year at Sandia National Laboratories, a large government laboratory devoted primarily to nuclear weapons design and energy research, working my first professional job out of graduate school. Sandia Labs sat in the desert south of Albuquerque, New Mexico, within the perimeter of Kirtland Air Force Base. It was both an austere and esthetic setting to work in and, unfortunately, I was finding the professional atmosphere at Sandia equally severe.

I am a plasma physicist, which means that I study matter in its fourth state—the one following solid, liquid, and gaseous states. In the plasma state, matter becomes incandescent and conducts electricity, like the aurora and the lightning bolt. Most of the matter in our universe is in the plasma state. Plasma is the "stuff" of stars and galaxies, so my scientific explorations ranged widely. However, my primary job at Sandia was a great deal more limited: the study of the flight of intense beams of high-energy electrons through the air. Sandia was trying to develop this technology as a weapon to shoot down missiles.

Learning the kind of lesson that lots of us learn on first jobs, I had become involved in a technical dispute with a senior scientist on the project. The quarrel had become a personal one, not helped by the fact that I had been proven technically accurate while at the same time, perhaps, choosing a poor strategy for making that point. While the Cold War with the Soviet was in its icy depths, I found myself in my own little cold war learning firsthand that science is a very human activity, tinged with passion as much as logic—my own as well as other's.

To divert myself from these problems, I had been studying the planet Mars and trying to write some science fiction. I found comfort in the idea of a future where the human race was alive and well, having survived the Cold War and evaded nuclear destruction, particularly since my job was unrelentingly about the potential for our failure.

In the red deserts and mountains surrounding Albuquerque, so different from where I'd grown up, I developed a sense of how it might be on Mars. I was thinking about using Mars as a setting for a fictional adventure for the human race in the future. So I had a real interest as I relaxed in front of the TV with my family, in the glow after a holiday dinner, waiting for what I thought would be something akin to a Flash Gordon spoof.

However, as the television segment began, the tone was surprisingly serious. Suddenly, the images of the "face" recovered by Vince DiPietro and Greg Molenaar flashed onto the screen. I was galvanized. My bemused smile vanished. This was no egg plant that looked like Richard Nixon. It looked startlingly like a human artifact, not unlike sculptures created by the Aztecs. It was stunning—and it triggered a dim memory of something else I had run across in my studies of the red planet. This was not the first feature discovered on the surface of Mars that resembled an artifact of a dead civilization. There were pyramids! Now just where had they been?

Then I remembered. While studying Mars I'd recently pored over Carl Sagan's excellent book *Cosmos*, in which he had briefly discussed the

Pyramids of Elysium, two fairly peculiar objects, on a part of Mars called Elysium Planum, that resembled small mountains carved to look like pyramids.[6] I remembered being somewhat annoyed with the reference since it had raised the question of intelligent life on a planet that appeared unable to support even primitive life. Like nearly all scientists following current developments in space, I knew the Viking life experiment had been negative. Mars was lifeless and had always been lifeless. That was the official conclusion. It was a simple story supported by good data. The question was closed. I had accepted this official line. Could I have been wrong?

I felt like the character in the *Star Trek* episode "The World is Hollow and I Have Touched the Sky," who lives inside an asteroid and discovers one day that what he has been told is the sky is really a solid roof. Like Galileo, after he had first seen the moons of Jupiter through his telescope— like beautiful pearls on a string—and known in an instant that the heavens were not as the Church had said, my mind opened. Could the accepted scientific picture of Mars as perpetually lifeless be inaccurate? I sat staring at the program trailer considering afresh the possibility of life on Mars and realized that I had to know for sure and discover the truth for myself this time.

According to the program two men had done this research: Vincent DiPietro and Greg Molenaar, who worked at Goddard Space Flight Center near Washington, D.C. I burned the names into my brain. I would call them as soon as I got back to work after the New Year's holiday.

Vince DiPietro and Greg Molenaar had gone public with their findings and images of the unusual feature on the Martian surface at a press conference in May of 1980. A public relations manager from NASA had consoled them in advance, "I hope you have better luck than we do." They didn't. Attendance was poor. Only a handful of interested souls attended: a couple of NASA astronomers, a science news magazine reporter, and a radio commentator. It was very disappointing.

Not everyone shared Vince and Greg's humble opinion of the public's response, however. To someone somewhere, Vince and Greg's press conference was an outrage. They had to be stopped and the heat made its way to their employer. This was the same employer who had, just a month before, given the two of them an award for their development of the SPIT process to enhance the Mars images. An inquisition convened.

Vince faced the confrontation with their bosses with fatalistic courage: he would go down fighting. Greg was worried about the outcome, though. His supervisor had demanded he erase all of the programs they had done together. Expecting a serious dressing down, Vince and Greg found they were being charged with misuse of government property: the NASA imaging equipment used to process the pictures. Then they were shown documents they had signed when they were hired forbidding them to carry out this kind of work. This was serious all right. Too serious.

Soon it was clear that this meeting was all about building a case for firing both of them. Only when there was a break in the presentation of the "evidence" against them did Vince ask to show the inquisitors the written permission he and Greg still had in their possession. Clearly, no one had expected this turn of events. After an icy pause, Vince and Greg were told they were free to go. There was nothing more to discuss. The meeting came to an end and they still had their jobs to return to, which they did in stunned victory. Suddenly everything was fine again, as if nothing had happened.

All the storming over Mars was nothing new, however. Years before, Mars itself had shown NASA how far from tranquil it could be. In Noachis, in the south of Mars, an enormous cloud of yellow dust suddenly appeared. It was soon followed by an explosion of reddish dust clouds. After brewing in Noachis, the storm broke loose and raged forth to the northeast, growing as it raced across the surface of the planet near the equator. With 500-mile-an-hour winds, it tore at the surface, lifting dust clouds like a wall reaching up to the heavens more than 13 miles into the thin Martian air.

In the control room at JPL there was no immediate concern as the dust storm appeared in the images from the Mariner 9 cameras that flashed on the screens that September 22 in 1971. With Mariner 9 still six and a half weeks out from Mars, there was plenty of time for the air to clear. Dust storms had been seen on Mars before—through telescopes. They always occurred near perihelion, the closest point in Mars' orbit to the Sun, when solar heating would reach its maximum. The storms would rage for a few weeks over parts of the planet and then die out. This was expected climatic behavior on Mars, a mere outburst of meteorological rage, just a part of the planet's bizarre character. Such events were predictable and planned for in the Mariner 9 mission. However, it soon became evident to the controllers at JPL that this dust storm was becoming much larger and more violent than any observed before. The storm's violence and massive growth was stupefying. It doubled and redoubled in size, spreading north and south. Within days it had encircled the planet and before the astonished eyes of the controllers at JPL, it invaded deeply into the polar regions to completely cover the planet, its violence unabated.

Two Soviet probes, Mars 2 and 3, arrived at Mars during the storm and haplessly tried to take pictures of the surface. Both of the craft had been preprogrammed and their now pointless missions could not be changed. Once their cameras had exposed the preprogrammed number of images, they shut down, despite the fact their pictures showed only dust. Both of the spacecraft released landers into the storm-wracked atmosphere with robotic compliance. One malfunctioned and the other crashed and returned only scraps of data. Mars had utterly defeated the Soviet missions.

The Mariner 9 fired its rockets to slip into orbit flawlessly. It was November 10, 1971, six weeks after the storm had begun, and even among the mission managers at JPL there was an atmosphere of crisis. From pole to pole the surface of Mars was obscured by roiling red dust clouds. Only four dark spots could be discerned on an otherwise blank disk of Mars. These spots were later identified as enormous volcanoes 13 miles in height—the only features on the surface tall enough to escape the cover of dust. As the spacecraft rode in orbit against the stars, skimming above the now turbulent and opaque clouds of red dust, the mission mangers waited and worried.

Mariner 9 was the first Mars orbiter. Its twin sister had ended up in the Atlantic. Anything might go wrong as the dust storms raged, preventing the completion of its primary mission: the global mapping and detailed examination of specified sites on the surface of Mars. But surely, they all agreed, the dust storms would abate soon, in a few days. Common-sense weather analysis from Earth told them that storm systems draw their energy from differences in pressure, temperature and moisture between various areas of a planet's surface. Surely, if a storm fills the whole atmosphere of the planet, the differences are all being mixed away. The storm must end when its energy source is cut off. Good theory. But still, Mars disobeyed the laws of the Earth and the dust continued to blow.

On and on, day after day, weeks stretching into months, the great storm raged. Red dust clouds were now thrown upward—even into the Martian ionosphere at the very edge of space, 45 miles up. At JPL the mission controllers went through every stage of anxiety, despair, and disbelief, as their spacecraft, still working flawlessly, circled hopefully above, snapping an occasional image of the extraordinary sight below. The number of pictures taken in the vain hope of finding a clear view to the surface grew.

On everyone's mind was the certain knowledge that if one major component aboard the craft failed, the spacecraft itself would fail—having seen nothing but dust clouds—as had Mars 2 and Mars 3. The Martian atmosphere had become one vast roiling mass, expending the energy of thousands of earthly hurricanes at once, the energy of millions of hydrogen bombs a second. Russet cyclones raced across the cloud tops. What energy source could drive this planet-wide engine of wind and dust? "When will the storms end?" demanded the JPL managers of the Mars atmospheric scientists, who could not answer. Finally, in resignation, the spacecraft was mothballed, going into "fetal position" in space in an effort to conserve the precious maneuvering gas it had been consuming with the constant adjustment of solar panels and the operation of thousands of moving mechanisms, all of which had been for nothing.

From then on Mariner 9 slept, drifting in orbit for months. Finally, in January, when the great dust storm, the largest storm seen on any planet in history, began to break up, the atmosphere began to clear. By March, the atmosphere had cleared completely, and the mission scientists at JPL feasted on the sight of a newly revealed planet with huge canyons and enormous volcanoes only guessed at before. There were no canals on Mars, but there were numerous vast dry river beds, telling us of running water that once coursed over hundreds of miles. When Mars finally revealed its surface features, it was a new, water-carved planet that we saw.

Sometimes grace comes in funny packages: discovering that Mars was once wet wasn't the critical information for Earth; it was the storm. Here we were, waiting impatiently for the storm on Mars to go away, but Mars was telling us that what we really needed to pay attention to was the storm itself. Mars was screaming out something critical to our survival. It's a good thing someone was listening—because most of us weren't even aware that there was any warning to be heard.

It happened something like this. A bunch of Mars boys were sitting around (boys never grow up completely, although in reality they were a group of prominent Mars scientists).[7] They were waiting for their latest and greatest spacecraft to take some pictures of Mars so they could experience a surge of adolescent elation in private and then publicly make some significant serious-sounding pronouncements about various aspects of Mars geology and climate, etc. However, there was a really big storm on Mars and the boys were suffering from various forms of delayed gratification and boredom. They were probably apprehensive, too. Everyone on the project had worked very hard, their government had spent millions, they'd gotten their spacecraft to cruise to a planet over 56 million miles away to exactly where it was supposed to be and then ... dust! So there wasn't really too much to do, other than watch this mother of all storms develop and think about things.

Now there are a lot of things they could have been thinking about: how to fill out progress reports for government auditors or what new job they would have to get if the spacecraft failed. But if you think they'd really be doing those things, you don't know Mars boys. Instead, they were obsessing about the storm and wondering why Mars should have such dreadfully annoying weather when Earth didn't. Or did we? After all, why should Mars have all the really bad weather? The Mars guys then started thinking about the serious dust storms that had occurred on Earth. And, knowing how these kinds of conversations go, it might have gone something like this:

"Hey, who says we can't have dust storms like that on Earth? We had those really bad dust storms back in the '30s!"

"You call those dust storms? Those were nothing . . . just wiped out everything in a few regions of North America for a couple of years. Who cares about that? We're looking for pole to pole here. What energy source could drive that?"

"What about Krakatoa in 1883? Now that was a dust storm!"

"Krakatoa? Big? Are you serious? When Tambora erupted in 1815 it put so much dust in the air that we didn't have summer in 1816."

"What about if we got hit by an asteroid or something? That could really kick up some dust!"

One thing led to another and the next thing you know the Mars boys, after all kinds of speculation and calculations, just couldn't conjure up a storm on the Earth that was as powerful as the one they were watching on Mars—a planet-wide storm powerful enough to end all life on Earth. They kept obsessing on the question as the storm kept swirling in front of them. "Why?" it seemed to demand. A volcano couldn't do it. An asteroid might do it but that was too unlikely. They knew it had to be something, but they could think of nothing that could raise millions of tons of dust high enough into the atmosphere to cause a planet-wide dust storm. Nothing could do it, they thought—until a few moments later, when deep down in the dark well of unconsciousness, a veil slowly dropped, releasing a hideous, terror-filled nightmare. It percolated slowly, inescapably upward and upward—building and billowing, like evil incarnate, into Earth's highest stratosphere in the dust-covered form of a giant, roiling nuclear mushroom.

Clearly the storm on Mars had not been caused by a nuclear explosion. Nevertheless, the inquiry about the storm on Mars had shown us something vitally important about our Earth. From that point, it became clear. Our planet could not survive the climatic changes brought on by a nuclear war. We could not survive a nuclear winter any more than the dinosaurs could survive the climatic changes brought on by a meteor strike. And while it took many years for the thought virus "nuclear winter" to infect us all, with the help of scientific journals and magazine articles, eventually this idea penetrated humanity deeply, moving even into the very places where the hearts and minds of those who were making bombs could see its truth, too.

A s I tucked my daughter into her bed that night after turning off the television, I considered what I had just seen—and the implications. What would happen if we discovered evidence of life, intelligent life, on Mars? To begin, I chuckled to myself, the Cold War would become untenable. Who could argue over whether Marx or Jefferson should be required reading in European schools with that kind of discovery flying in from the sidelines? It would be like a barnyard brawl that ends when a nearby barn catches on fire. We might even work together on that project, I thought, thinking of the scientific zeal that could be unleashed. I collected my books on Mars and thumbed through them until I found images from the Mariner 9 mission, including one of the Pyramids of Elysium.[8] I read again about the extinct water channels. It was clear, Mars had once been much more Earth-like, with rivers of water on its surface. The mere presence of liquid water meant that both the temperature and atmospheric pressure had been Earth-like at one time. But that was supposedly over three billion years ago. It also seemed clear to me instantly, that if the "face" was what it appeared to be, Mars must have had a climate like Earth for much longer than anyone suspected.

Back at work next morning I stood with a group of other scientists in the hall, drinking coffee and swapping holiday stories. I enjoyed the feeling of camaraderie with the other staff members at Sandia who were a superbly talented group. I mentioned the "Face on Mars" piece to the group and was immediately greeted with annoyed disbelief. Fortunately, a woman physicist who had also seen the program rescued me. I learned from this experience that even among highly educated and astute scientists and engineers, the possibility of a humanoid artifact on Mars elicited quick emotional reactions, including fear. However, once the other scientists heard from two witnesses that the feature, whatever it was, really existed, annoyance turned to curiosity and a general agreement that it would be a stupendous find if it was an artifact.

These were days of the depths of the Cold War with the Soviet Union, the Reagan arms buildup, the endless war in Afghanistan, the shooting down of Korean Airlines flight 007, and a renewed examination of the consequences of a nuclear war. I was quite dedicated to the defense of the United States, and was a fervent anti-Communist, especially after visiting Russia in 1972. But at the same time I was dismayed at the seemingly endless character of the Cold War. Confrontation between the United States and the Soviet Union kept the whole human race in peril of nuclear destruction, perhaps by accident.

In my job as a physicist at Sandia, I was working on Directed Energy Weapons and fusion-related projects. I was on the scientific front line of the Cold War. So were all my cohorts in those days of "Star Wars" defense research.

Star Wars, as it was known popularly, or Strategic Defense Initiative (SDI), as it was known officially, was in its initial stages then. The Reagan announcement of SDI had produced delirious excitement throughout the laboratory and the whole of the defense research establishment, since it seemed to promise an alternative to the traditional nuclear arms race we had been locked into for so long. It lifted a pall of depression that had loomed over all of us involved in defense research. Curiously, that pall of depression had had its origin in the terrible storm over the plains of Mars.

The nuclear winter scenario described by Carl Sagan, Richard Turco, Owen Toon, Thomas Ackerman and James Pollack was very disturbing to my colleagues at Sandia, and to me too, though I had never admitted it. Though it was never acknowledged publicly, indeed it could not be, morale among defense researchers plummeted in the wake of the publication of the nuclear winter studies, most notably those by Carl Sagan in *Parade*. In the United States, at least, scientists read their Sunday newspaper inserts. In the defense research establishment, particularly in the nuclear weapon design groups, the nuclear winter research confirmed that the environmental consequences of nuclear war between the superpowers would not be survivable, even if the initial blast and radioactive fallout could be escaped. The reason nuclear winter appeared possible was that it had been found that the injection of vast amounts of dust into the atmosphere of a planet, as in a nuclear war, could result in a semipermanent global state where the dust remained turbulently suspended in the atmosphere, cutting off most sunlight to the surface. Thus, the survivors of a nuclear war's initial effects would most probably starve and freeze to death, in a second chapter of the war that would take more lives than the first. It was this concept that had settled over the Sandia labs like the shadow of doom.

By having taken the wind out of the sails of the arms race, and demonstrating that nuclear war wasn't survivable, Mars, indeed, may have saved the Earth. While we like to think of Mars as remote and disconnected from our world, the truth is that we share more than a past. We share a future.

Human fascination with Mars is understandable, and even necessary, particularly when you consider that humanity will most probably be living on Mars a billion years from now. While this is a nearly incomprehensible length of time—approximately 333 times the length of the entire history of humanity—eventually it will be much too hot to stay on Earth, and Mars is our most likely next home world.

As our Sun, now 4.5 billion years old, ages, it gets hotter, and as that happens, the narrow temperature zone for human habitation, in which water remains an unfrozen liquid without boiling off altogether, gradually moves out beyond the Earth. (A small minority of scientists claim that this is a factor in global warming today.) Eventually, a billion years from now, Earth will be blazing hot and bone dry. Like Venus, someday Earth will no longer be within the habitable zone around our Sun. And Mars? Because a brighter Sun will send its warming energy further out into our planetary system, Mars, instead of being a frozen wasteland, will be a sunny, warm planet, safely orbiting the Sun well inside the outwardly moving habitable temperature zone.[9]

Although small by Earth standards, Mars could be an ideal place for human habitation someday, provided it is "terraformed", or made Earth-like.

According to some, this process may be achievable over a period of 10,000 years or so, which would require the participation of many generations of humans working to prepare for a future that they would not live to see. In addition to the ability of people to work together toward a very long-term goal, the achievement of a Mars that is habitable for plants, animals and humans will require a number of significant advancements in human development and evolution. The most obvious are the skills and technological advances required for economical long-distance space flight. The less obvious are the development and evolution of ways to maintain our own Earth over the next 10,000 years and more until Mars becomes habitable.

While making another planet Earth-like seems an outlandish notion, perhaps more science fiction than serious science, at one point, not so long ago, the idea of "flying" was strictly fiction and any serious notion about traveling in space to other planets was considered a grave disorder of mind. The future frequently looks like fiction. That's one of the reasons it is so vital for humans to be able to imagine—to see powerfully with our own creative vision—wondrous, extraordinary possibilities for the future.

When our Sun gets brighter, Mars, instead of being a frozen wasteland, will be, more or less, a temperate wasteland, without flowing water, a breathable atmosphere or familiar looking plant-life. However, if we wanted to begin the process of terraforming Mars before our Earth is blazing hot, we might begin by warming Mars, in much the same way that we are unintentionally warming the Earth today. Creating a "green house" effect by building up atmospheric gases would be desirable on Mars, but initially we wouldn't have the available atmospheric oxygen to burn things, which is a major method used now on Earth. Perhaps a small fusion device could be used to help generate greenhouse gases; water vapor, for example (which is a very good greenhouse gas, in that it is both non-toxic to humans and effective in holding in the heat from the Sun) could perhaps be produced from the vast quantities of underground water that Mars shows indications of hiding.

The northern polar cap of Mars consists primarily of frozen water. It currently reflects about 77 percent of all sunlight directed its way. If we darkened it a bit, the polar cap would absorb some of the heat from the Sun that is currently being reflected as light. That heat would begin to release water vapor. Perhaps we could find a blue green alga that grows in cold climates and spread it on the polar ice of Mars to darken it, as Johannes von Buttlar, the famous German Mars science writer, suggests.[10] We might have to bring algae from Earth, but there's a "non-zero" chance we could find some living right on Mars, too. Not only would algae absorb more of the solar light and contribute to heating, they could also begin the process of photosynthesis, consuming carbon dioxide and releasing oxygen.

Over time, as the processes of atmosphere building and warming continue, the new, denser Martian atmosphere would increase in barometric pressure. Eventually, Mars would have warm temperatures, clouds, rain, and water that flowed on the surface. However, in order to have adequate oxygen for humans, we would need to grow lots of vegetation. Also, both humans and plants need protection from the ultraviolet rays of the Sun, which means that we would have to build a protective ozone layer around the planet.

While the ozone layer was being built, we might begin to live on Mars inside domed centers or stations constructed with materials that we are able to find there. For example, there are two newly discovered craters in the Cydonia region that appear to have bottoms filled with water ice.[11] These would be likely places to build domes. Building domes inside craters would automatically provide protection from the elements (especially the cyclone-like Martian winds), reduce the need for materials and excavation, and help contain the newly released atmosphere. Once the atmosphere beneath the domes warmed slightly, the ice would begin to melt, providing essential water which would then be a source for the chemical reactions needed to release atmospheric gases such as oxygen. Like a terrarium, a domed crater on the surface of Mars could become like a tropical oasis, complete with water for recreation.

So this is how it might begin. Given adequate time and ingenuity, it is likely that all aspects of terraforming Mars can be worked out. Given human will and imagination, it is likely that future generations will unfold this, or some similar, scenario on Mars. Nevertheless, we must evacuate Earth as an abode of life someday, unless we can figure out how to keep cool inside what will otherwise be an uninhabitable zone, or move our planet to a new location. Of course, we may be making trans-dimensional jumps to other galaxies for the weekend a billion years from now. Who knows? Arthur Clarke's point that a sufficiently advanced technology is indistinguishable from magic is certainly worth considering when we imagine the future we truly want. No matter what the current circumstances, we are at liberty to envision the future as powerfully as we wish, including aspects that may seem highly improbable to us today. Our future is not limited to what we now think is possible, nor is it limited by what now seems impossible. In the future, anything is possible. However, no matter what future scenario we enact, part of our future will more than likely entail looking beyond Earth as the only abode for human life.

So, we may be terraforming Mars someday—but only if we can master the principles, science, technology, and wisdom for managing the planetary systems required, while we are still right here on Earth. And Mars, in its unfathomable generosity and wisdom, can perhaps teach us how.

The best minds of the Mars atmospheric community pieced together what had happened on Mars to create the global dust storm. The energy source was apparently solar heating of dusty air, and the convection was vertical. Warm air rose and lifted dust, which absorbed more sunlight, heating the air further, so that it rose even higher before it cooled. Surrounding air spiraled in at low level to replace the heated air and raised more dust, thus repeating and intensifying the cycle. The "dusty hurricane" model gave a rough model of what was seen, though large spiral structures were seldom apparent. The storm was not a front moving from hot to cool or vice versa; it displayed vertical convection mode—a pattern of up and down drafts—and it could be very long lived. During this storm, the model showed that sunlight would be completely absent several miles above the Martian surface, everywhere but the poles. It was also concluded that the results of this model could happen on Earth, too, given that enough dust, ashes and smoke were injected into the atmosphere at once.

So this last echo of the thunderous winds of Mars had erased all doubts among the family men laboring at Sandia. When they said that the nuclear war strategy of the United States was Mutual Assured Destruction, or MAD as it was called, many of these men had not followed this idea back to the doorsteps of their family homes. Many scientists working in the government nuclear weapons complex, of which Sandia Labs was a part, secretly felt that they would survive a nuclear war, along with their families. I personally had always thought this was ridiculous; however, that was just my way of not thinking about it. For many others, it seemed they believed they were keepers of the magic flame, and since they understood it and even worked with it, it would not harm them. Then the nuclear winter studies shattered that illusion forever.

Now the scientists at Sandia had no choice, they had to confront what MAD meant to them personally. I had resolved long before this that I did not want to survive a nuclear war, and that there was no such thing as a limited nuclear war, or even tactical nuclear weapons. There was only total nuclear war, and MAD with its various means of igniting it. I had heard that in realistic war games and simulations of limited nuclear war, communications had disappeared very quickly, and so all commanders had tended to fire every weapon they had before they were destroyed on the ground. I was also aware, long before the nuclear winter, that a nuclear war would destroy the ozone layer for several months or years. Therefore, the deep depression that seemed to sweep over Sandia did not affect me as severely as it did others.

"You know, John, I used to think if I knew there was going to be a nuclear war, I'd get the wife and kids into our van and head for the hills," explained my officemate Malcolm, one afternoon. He was normally a very genial fellow. Astonished, I turned around from my desk to look at him.

"Now," he continued, "my plan is to get up on the roof with a six pack of beer, and watch the whole thing go." I looked at him blankly. Finally I mustered enough thought to say something in response. "Malcolm, what kind of beer would you be drinkin'?" I asked. He chuckled and returned to his work and I to mine. I must have heard three different versions of this story that spring of 1982—all beginning "I used to think if I knew there was going to be a nuclear war, I'd get the wife and kids into our van and head for the hills but now . . ."

I knew that this feeling was not confined to our side of the Iron Curtain. I stayed up late one night to watch *Letters From a Dead Man*, a horrifying Russian film about the aftermath of a nuclear war. As I was discussing the film with a colleague in the hall one morning, he asked me what the movie was about. "It was about life after a nuclear war," I replied.

One of my supervisors overheard this as he walked by and interjected loudly, "There is none." He kept walking, shutting the door to his office behind him. End of conversation.

CHAPTER 5

Bolder and Boulder

I N MY OPINION, THIS IS AN ARTIFICIAL OBJECT . . . I typed this
message furiously on the broken-down phone terminal that I'd
borrowed from a friend back in the days before modems were
commonplace. The terminal was not what I had hoped for. It printed only
capital letters, making every message look as though I were shouting, and
its cooling fan was broken, so I was forced to improvise one from a toy
motor with a propeller and a small transformer that plugged into the wall. I
could only type or receive for about forty-five minutes before it would start
to overheat, often just as a crucial message was being sent or received. Yet,
despite the frustrations, I was happy.

While at first I hesitated to become involved with the Mars investigation,
I also felt the work demanded scientific attention and I was inspired by my
image of a true scientist—a committed individual, someone who had a
deep concern for the impact of science on humanity, who was also a bit of
a swashbuckler as well—someone like my uncle, William Siri.

Like many physicists involved in the Manhattan project in the Second
World War, he had carried away from that experience an intense desire to
conduct true science which would serve, rather than endanger, humanity.

My uncle had gone to the ends of the Earth in search of that true
science. He went to Antarctica, riding in snocats across the blue ice. The
snocats carried special pontoons which extended ahead to detect the
crevasses that could swallow them whole. He climbed Mount Everest on an
expedition in 1963, serving as second in command and going most of the
way to the top. He carried a centrifuge nearly to the summit and gathered

physiological data as the climb progressed. He made the discovery that on such a mountain, climbers' blood was much thicker than normal, in part because everyone was constantly dehydrated. He carried back with him an almost spiritual awe of the sight of the surrounding Himalayas, and brought back a crystallized clarity about how limited and fragile the Earth was. He applied this knowing by serving as president of the Sierra Club from 1964 through 1967, and his visits to our house when I was young always inspired me, as he would patiently discuss everything from unified field theories to the then fledgling environmental sciences.

Inspired, I took up the gauntlet of Mars, feeling that to ignore such a possible great discovery at Cydonia would be shirking my duty. To ignore such a discovery, one that might alter forever the human perspective of humankind and the cosmos at such a perilous time in human history, seemed unconscionable. My uncle would have asked no less.

So I found myself in the middle of the dialogue, the investigation, square at the center of the battle for an important cause, as I had always hoped. From my very first conversation with Vince DiPietro, I was suddenly part of the investigation of Mars, joining an informal group of scientists and technologists in the pursuit. The "face" was no longer the only subject of interest in Cydonia—the whole question of Mars' past was of interest. The more we looked, the stranger it became. Another large pyramidal-shaped form, similar to those in the Elysium Planum region that Carl Sagan had talked about, had been discovered in the Viking images of Cydonia. Greg and Vince had done a great job, the data were sound, and we spent two months in analysis and discussion, slowly edging toward an audacious conclusion. Our excitement mounted. This, I thought, is where I really belong. As the music of MTV, the new music phenomenon that our three-year-old had accidentally discovered while turning channels, played in the background, I typed away. Our exploration of Mars was carried out to the accompaniment of the Police, the Eurythmics and Tommy Two-Tone.

The Cold War swirled around us as we worked through the winter of 1984. The U.S., still numb from the massive truck bombing of the Marines in Beirut, was now being pulled deeper and deeper into the civil war in Lebanon, under the mistaken belief that, with its kaleidoscope of changing alliances and factions, it had something to do with the balance of East–West power in the region and, hence, the Cold War, when in reality it was nothing that rational. The Lebanese Druse, Shiites, Christians, Sunnis, and Palestinians, together with the Syrians, Israelis, and other combatants in the region, were merely continuing a cycle of revenge and atrocity predating even the religions these groups wear like armor. The irony was that while all of us were aghast and shook our heads at the horrors of Beirut, the one-time "Jewel of the Levant," we ourselves were engaged with the Russians in a struggle that, in thirty minutes, could make every city on two continents look like Beirut's nightmare. The greater structure of the Cold War was hardly more rational. I viewed Beirut as the world in microcosm.

The intensification of the Cold War had spawned a world-girdling string of armed conflicts. It seemed at times that the United States was continually in some state of undeclared war. One day 200 Marines were killed in Beirut; the next day U.S. paratroops were landing in Grenada and fighting the Cubans. Two thousand miles to the west, the civil wars in El Salvador and Nicaragua raged on. The refugees from these conflicts streamed north to the United States, some fleeing the right-wing death squads in El Salvador, some fleeing the East German-trained Nicaraguan secret police.

While at work I felt the Cold War as an abstraction; in the rest of my life it was inescapable and real. In the church my wife and I attended, St. Andrews Presbyterian, half the congregation was on the faculty at the University of New Mexico and half worked at Sandia Labs or the Air Force Weapons Lab. This made for a highly aware and highly divided congregation. We were also painfully sensitized to the plight of those Central American refugees streaming across the border from old Mexico into New Mexico, in addition to the numbers of Mexican citizens trying to escape the poverty of that nation.

The Reagan administration considered all of these people "economic refugees" and would round up those seeking remedy for political persecution along with everyone else who had entered the U.S. illegally and put them in detention centers. Ultimately, the aim was to fly them back to their countries of origin, even if it was into the waiting arms of the right-wing death squads of El Salvador or the Sandinistas' forces. However at odds with my job it must have seemed, I felt this policy was utterly heartless and indefensible, even though I was a Reagan supporter. My liberal wife was not content with outrage, she was moved to action. She joined the "Sanctuary Movement," which offered the traditional idea of church sanctuary to the refugees. She volunteered to lead a completely legal effort by the church to sponsor such refugees and get them out of detention centers and help them get jobs, a place to live and then help them petition for political asylum in Canada. Soon she succeeded and we had a house guest: Roberto.

Roberto was an exceptionally bright and disciplined young man, the son of a middle-class family. He was fleeing the Sandinistas, who wanted him for their army to fight the Contras. To Roberto, service in the Sandinista Army appeared as welcome as wartime service does for anyone not supportive of a government: like pointless death. One of his friends had been taken away to serve in the Sandinista Army, was reported killed, and returned home in a coffin. But when the boy's family opened the coffin, they made the grim discovery that it did not contain their son's body, but rather that of someone else. This awkward situation was resolved, much to the parents' horror, by posting armed soldiers over the coffin to ensure it was buried without being reopened.

After being smuggled out of Nicaragua by his family, Roberto made his way through the war-torn landscapes of Central America and across Mexico, along the Pan-American Highway over jungle and desert, following the path of shadows cast by so many other refugees. Ultimately, he arrived on our doorstep seeking sanctuary, but the trip had been arduous and risky.

Roberto did not drink or smoke and was a perfect house guest. However, as an example of the confusing ideologies of the times, some of our more liberal church members expressed dissatisfaction with Roberto's politics and his status as a dodger of the draft of a socialist country. I reproached myself for pushing something as abstract as Mars studies when the world seemed to be spinning toward catastrophe, and so I resolved to work within my church to take some concrete action against the mounting intensity of the Cold War. I became involved with several other church members trying get a resolution passed by the Presbyterian Assembly to urge the renunciation of the option of "first use" of nuclear weapons.

Renouncing "first use" was an idea first proposed by McGeorge Bundy, an old Johnson administration hand, and well respected. "Mc Bundy," as he was known, pointed out that the doctrine of first use of nuclear weapons by the United States had been formulated during the Eisenhower administration in the early '50s when the Soviets had no nuclear weapons to speak of. However, by 1984, when the Soviets had them in abundance, first use became tactically unwise and strategically suicidal. Was it logical to begin a nuclear war to prevent the loss of West Germany when it would in fact devastate Germany and the whole Earth as well? However, logic and nuclear weapons, I was to find, were frequently connected in ways I did not understand. Although I continued to pursue the Mars investigation, which at least had the advantage of making logical sense to me, I sandwiched it between work on my no-first-use committee and helping Faye with her refugee.

We were young and full of energy then, consumed by a passion to do something about a situation far larger than ourselves, a horrifying conflict that had to be escaped. At the same time, that conflict provided me with a living. So, by day I stoked its scientific boilers by pondering the paths of electrons, only to return home each night to where part of the human cost of the Cold War slept peacefully in the room down the hallway.

Humans have resolved many thorny issues that initially seemed virtually unresolvable for years or even decades. Yet, even when there is no straightforward resolution available—even when the discrepancies mean that we act by day in service to one objective and by night to its antithesis— it is better to have our eyes open and be willing to take in the paradoxes. Not everything pans out nicely into neat categories of "right" and "wrong" in real life. But, when we stare down the truth, no matter how much we dislike it, we stand a much better chance of ultimately resolving serious problems than if we simply deny them.

Earlier we talked about how humans are "blind" to some of the things that endanger us because those threats are invisible. We simply cannot perceive some threats. However, the inability to physically see is not the only form of blindness that humans sustain. There are others. One of the more potentially damaging of these is the ability we have to deny what is occurring. If for some reason a circumstance is perceived as too threatening, some people "switch on" a psychological coping mechanism which allows them to deny—or be "blind" to—actual circumstances as if they didn't exist. How many times have we heard the spouse, whose unhappy mate was cheating, lament, "I should have known. It was going on right in front of my eyes, but I just couldn't see it."

Humans do strange things when the stakes are high. When facing the stark reality of a situation might mean major emotional pain or physical and financial distruption, one coping mechanism is to unconsciously push away evidence in an attempt to gain some measure of protection. It's a great trick we pull—using the selectivity of our own neo-cortexes to "lie under oath" to ourselves. In fact, perhaps it does provide some sort of psychological buffer by slowing down the rate of dawning recognition. In any case, this is either a normal human trait and part of our physiology, or it is a learned trait and very common in Western culture. It is common, yet rarely do we consider that the people we may look to for guidance—such as scientists, political leaders, policy makers, and media representatives—are themselves capable of denial.

While most of us are aware that denial is a factor in circumstances such as marital infidelity, drug addiction, alcohol abuse or gambling, there is a much subtler form of denial that we sometimes fall into. In this version we do not deny the truth of the circumstance, but rather, we seek to immediately put some distance between the circumstance and any impact it may have on us.

In "distancing," as we'll call this mental process, we surround ourselves with a margin of protection. Our unconscious objective is to sufficiently distance ourselves from the problem so that it lies outside our margin of protection or "safety zone," leaving us with the perception of safety from harm. Here is an example of how distancing works.

"Although my mother died of breast cancer and my younger sister is a breast cancer survivor, I personally think of breast cancer as an environmental disease, preferring to ignore my clear genetic vulnerability." The irony is that much of the rest of the world feels a sense of reassurance knowing that genetics play a part in breast cancer. To them its genetic basis seems to offer some sort of psychological protection from the stark reality that the environment is a major contributor to increasing rates of breast cancer, and that many women with no relevant genetic background also get this dreadful disease.

Can you see that every woman has a choice of perspectives which help to distance herself from the threat of breast cancer? If our personal genetics make the genetic aspects of the disease seem too threatening, we have the option of focusing on the environmental causes. If the environmental causes seem to hit closer to home, then there is more "distance" in latching on to genetics as the significant factor. We are selective in the facts we promote as the "significant" factors.

And we are not alone in doing this. We get help. Those who care about us also do it in an effort to help us create "distance" from threats. "I told my husband that I was worried about the environmental causes of breast cancer and he reminded me that genetics play a big part." However, he might just as easily have suggested the opposite if that was where the greater distance was to be found. "I told my husband that I'm concerned about getting breast cancer like my sister did and he reminded me that the environment plays a big part, too."

In this example there is validity in both positions. It isn't so much about whether something is true or not, but whether the particular fact we latch onto does or doesn't create distance. The point is that we disassociate ourselves from potential threats, even when they may actually be very relevant to us. We frequently behave as if the full truth is too bitter to face, so we quickly seek contradictory evidence, no matter how shaky. Have we become so addicted to feeling good that we are doing so at the expense of being honest and effective? Are we so threatened by uncomfortable emotional feelings that we would allow real threats to persist in order to avoid feeling badly about them?[1]

While all sorts of strategies can create distancing, the underlying fear-ridden motivation driving us to seek distance is seldom expressed or even consciously acknowledged. One strategy we use is to push the threat further away in time. We hear someone say, "Too bad about those floods in the Midwest." (Our unexpressed fear might be: We are overdue for a flood.) Our response might be, "We're lucky. We haven't had a flood here in decades." Physical distance can also make us feel safer. "I hear there are going to be more storms and hurricanes." (Fear: We might have a tornado here.) "Thank goodness we don't get hurricanes this far inland." Emphasizing our financial advantages can distance us from the problem. "I can't believe how many people lost their homes in that disaster in Central America." (Fear: It would be terrible to lose your home.) "Of course, the construction methods they use there are so shoddy, no wonder their buildings all got wiped out." Even cultural differences can provide distance. "You know, they had hundreds of fires in that country last year." (Fear: Fires, burning out of control.) "Too bad those people are still using slash and burn agriculture." Sometimes we emphasize tribal or religious differences. "I've heard that they are sending the Jews to work camps." (Fear: I might be next.) "I'm glad that we're Catholic."

Distancing is usually accomplished by a quick assessment in which we latch on to one fact, distinction or interpretation that allows us to "dis-identify" with the circumstance. Once we no longer identify with the threat, we gain some perceived psychological protection because of the distance between "it" and us. Our reaction is like trying to get away from someone with a bad cold: no exposure, no contamination. However, distancing is not so much a calculated response as a reflex—and it is a very widely practiced habit. See if you can catch yourself and others in the act.

Distancing has a number of advantages over denial. For one, we don't have to be completely blind to the reality of the circumstance. We can pretend we're being rational and dealing with reality, and we are—just not very deeply; nor are we examining how the truth really might apply to us or impact the world we live in. We are being selective in considering only those facts which place harm farther away. So we walk through the world with gauze-shrouded eyes and hands—so as to feel less pain—until we discover too late that we've already walked into a fire.

There is one more major implication of denial and distancing. One of the classic circumstances where humans distance or use denial as a psychological coping strategy is during the process of dying. Dr. Elizabeth Kübler Ross' book, *On Death and Dying*, clearly desribes how denial is often a stage people go through in coming to terms with death. Some people deny that they are terminally ill even to the very end. While most people do come to accept their own death at some point, Kübler Ross is unendingly generous in allowing dying patients to deal with the facts on their own terms, even if that means playing along with a person who needs to pretend everything is okay when clearly it is not.

Facing facts is not a survival issue for people who are dying. However, for others, denial of unwanted reality is potentially a very dangerous problem. Unless we can squarely face unwanted reality, we will be unable to effectively change matters. Ultimately, both denial and distancing are bankrupt strategies, since they are designed to make us feel better psychologically while keeping us from considering solutions and taking effective actions that could achieve successful remedies.

If we are to take our next step in conscious evolution, we will have to consciously allow stark truths to penetrate and exist in our awareness, even those which are potentially threatening, even those we are clueless about how to resolve. As challenging as that may be—and as sobering—facing facts is the first step to resolving them. We just won't call the fire department until we admit the house is on fire.

As we begin to face facts we free up the energy we had previously used to suppress them. We open ourselves to fresh possiblities and creative solutions that weren't an option while we were denying or distancing. In fact, much of the numbness, angst, resignation and cynicism that we feel is the result of endlessly suppressing how worried and upset we are about our circumstances on Earth. When we start to face the truth we may regain our "ability to respond," which is the real meaning of "responsibility." We may feel bad for a while when we again begin to face the facts, just like the wife whose husband has cheated is going to feel bad when she finally comes to terms with that truth. She may have to deal with guilt about her part in things, with remorse and regrets, with fear about the future, and with rage. But doing so will allow her to make conscious choices about what she wants,

to assess what's important now and to take positive steps. It will also permit her to begin to rebuild that relationship or form another one, and as importantly, rebuild her relationship with herself.

This, too, is how we can face the reality of our damaged environment. When we can see fully what has happened, we can begin to reverse the damage and heal our relationship with our own planet—the one that sustains not only us, but everything we love as well.

M ars consumed me. I was transfixed by the articles I read about the scientific results of the Viking expedition. On its strange, Moon-like surface, between the craters and within them, were dried river beds. It was the saddest sight on Mars, because it told a poignant and unmistakable tale. In that instant I grasped that Mars told an irrevocable tale of disaster.

What looked like dried river beds, like the arroyos, so common in New Mexico, meant there must once have been liquid water. Water is a combination of hydrogen, the most abundant element in the cosmos, and oxygen—which is the third most abundant element, after helium. After stars have burned up most of their hydrogen into helium, they begin to burn the helium into oxygen. With stars making abundant oxygen, the oxygen is available to combine with the hydrogen, which is also abundant. So it follows that water is the most common chemical compound in the cosmos, far surpassing ammonia, methane, and carbon dioxide (its chemical cousins).

Water is not only the most probable substance to find on or near the surface of Mars, it is also the only compound likely to have been in a liquid state under past Martian climatic conditions. Rivers indicated Earth-like conditions, since the presence of liquid water on Earth's surface basically defines its conditions of pressure and temperature.

Water could not flow 20 feet on the surface of Mars under present conditions. It would freeze, damming its own flow with ice, as it does sometimes on the far northern tundra of Earth. Yet, on old Mars, rivers had once flowed for hundreds of miles. So Mars' past was much different from its present. But how long ago was that warmer, wet past? The craters hinted at an answer.

The fact that Moon-like craters and Earth-like rivers co-existed, intertwined in the Martian landscape, meant that the surface conditions on Mars had once been like those on Earth, but an Earth with a much higher rate of cratering. Most of the objects forming craters on Earth are rocks—pieces of the asteroid belt—as opposed to comets, which are mostly ice. Mars is at the inner edge of the asteroid belt, which means that it has always been pummeled by crater-causing objects—far more often than Earth has. Its higher rate of bombardment means that craters would form faster than Earth-like erosion could wipe them out. On Mars, the violence of nature was continually unleashed. Explosions dwarfing human nuclear tests continually pockmarked its surface.

So, on the one hand, if we knew the rate at which craters formed on Mars, we could count the number of craters per square mile in the regions where the arroyos were and then estimate their ages. However, it became apparent from the scientific literature that we didn't know the cratering rate on Mars with any accuracy, so the age of the arroyos could only be crudely estimated. That fact in itself was interesting. On the other hand, if we didn't know the age of the arroyos, then we didn't know how long ago the rivers had stopped flowing, hence we didn't know how long ago Mars had been like Earth. Although we knew it had been like Earth for some period of time, we were in a poor position to make any further arguments about what had or hadn't occurred on Mars. "That was it!" I suddenly realized. That was the crux of the issue with the Mars "face," too. We weren't able to argue adequately either way—whether it was artificial or not. When we viewed it in the same manner as we did the dried river beds, suddenly the "face" made an odd sort of sense that it hadn't before. Clearly we were dealing with possibilities that could not be ruled out.

Nightly I debated these points and others with the members of the group, which by now included anthropologist Dr. Randy Pozos, Dr. Lambert Dolphin, a physicist at SRI, and writer Richard Hoagland, in addition to Vince DiPietro and Greg Molenaar. Eventually we all fully grasped what the evidence pointed to: a mind-numbing catastrophe on Mars. That a planet could have had a long period of Earth-like climate, with rivers and lakes, and, most likely, life—microbial and, perhaps, even advanced life—and then be transformed into a lunar wasteland spoke of ultimate disaster. What had happened on Mars? This question was particularly unnerving as we watched human folly run rampant on Earth around us. As Lambert Dolphin put it so well: "We felt were not looking at Mars, but Earth, some few years in the future."

New sites, potentially archeological, were also found on Mars. More and more, Cydonia was looking like the site of some strange off-world habitat—not quite fully geological, not quite fully architectural. Perhaps a combination of the two? How could we know? How could we be certain either way? In a place called Dueteronilus, a very strange sight appeared on a Viking image: a pyramid-like object perched on the debris apron of a large crater, casting a shadow 6 miles long. The object was the tallest for hundreds of miles, and had obviously been created after the crater formed in order to sit unscathed on its rim. Could a huge pyramidal mountain emerge on the edge of a crater after the crater was formed? Totally unlike anything we had seen with Earth's geology, this object was mind bending. Artificiality had to be considered. But that wasn't all. Near this new pyramidal object, the surface of a hillside was terraced like a rice paddy, and to the immediate south of the "face" more strange landforms were discovered—ramps apparently leading up to the tops of mesas and other strange bisymmetrical features.

The Independent Mars Investigation Team, as we were calling ourselves, was looking at these Mars anomalies more closely, looking with all of our combined scientific training. Using the most conservative explanations, we were trying to make sense of what we were seeing, and hoping for a rational explanation that would resolve our persistent questions. But instead of melting away under the bright light of our combined scrutiny, the question of artificiality was becoming more profound and persistent, and the implications more fascinating and terrifying. Curiosity gave way to transfixed dread.

We were like a band of engineers sent out to look at cracks in a large dam and finding gaping fissures larger than our worst fears—and then realizing that we are standing right at the bottom of the dam. Bad as that was, considering the wider implications of what we were seeing on Mars was even worse. We were like dwellers in a tinder-dry forest, one of two groups of avowed enemies, each roaming fully armed with flame throwers. Then, in our wanderings, on the very next mountainside, we discover the clear signs of a vast fiery inferno that had wiped out all life, totally and forever. We dwellers on Earth were playing a very dangerous game.

When a few infinitesimal molecules of the scent of smoke touch the olfactory tissues of a mammal, that animal (including a human) will spring into action no matter how deeply it might have been sleeping. Fire awakens a primal fear of death, triggering the flight response in birds and animals, who will use their wings or legs to fly or run away, leaving their nests and burrows in order to survive.

Trees can't run. A tree or plant must stay, attached to the ground, as the flames roar toward it, as its neighbors die, as its bark begins to warm, then sizzle, as flames explode up its sides and blast into its needles, as it is consumed in an inferno, giving its life-stuff back to the cycle of carbon.

There is a moment in a forest fire when a tree actually dies. One minute it is a living, growing being, the next, in a superheated flash, it is a black negative of its former self, standing silhouetted against a blazing aura of undulating flame—silver white on the inside, radiating outward in flowing bands of cobalt, orange and yellow—a tragically splendid, blazing torch. By the following day, there is nothing left of the tree but a few shovels of dead and dying embers and a thick layer of ash on the ground awaiting a cycle of regrowth. The rest has rejoined the air as particles of smoke and carbon dioxide, adding to the greenhouse effect and amplifying the problems made by humans. Global warming has claimed another victory.

While many of us are aware of endangered tropical rainforests and the critical part they play in the carbon cycle, not as many of us are aware of the threat to the northern boreal forests which are found in Russia, Canada, the United States, Scandinavia, parts of the Korean peninsula, China, Mongolia, and Japan.[2] They are particularly threatened because they thrive in a relatively narrow band of temperature range.

The boreal forests of the northern hemisphere have lived in a beautiful harmony with the Earth for tens of thousands of years. If you could speed up the process so that you could watch the ebb and flow of the forests' growth and movement over the last 18,000 years (since the time when the last major ice age began to recede), you would see that slowly the forests do "travel" in response to the movements of glaciers and changes in global climate. However, the forests can only accommodate very gradual changes, slower than the climatic changes that are being introduced today.

Like trees and plants everywhere, the firs, pines, larch and spruce of the boreal forests use carbon dioxide in the process of photosynthesis. Plants take in water and carbon dioxide, and in combination with sunlight, process these to produce oxygen, water, and food. During most of their cycle of growth, plants take in much more carbon dioxide than they release and are thus one of the major "sinks" or methods of eliminating carbon from the atmosphere. When they decay or burn, the carbon from their cells is released back into the earth or into the air.

The ability of plants and trees to process carbon depends upon several factors—available water, sunlight, their age, and the surrounding temperatures. A mature plant or tree, or one under stress from too little water or too much heat, will process less carbon dioxide than a young, healthy one.

About a fifth of boreal forests grow in areas at the edges of lakes or fens. These wet areas are covered with deep layers of decomposing vegetable matter known as peat moss, also a major carbon sink. In fact, it is precisely this situation—deep deposits of buried peat, transformed by time, heat, and pressure—that led to the formation of coal and oil. Most of the fossil fuels we burn today were formed some time between the late Cretaceous Period, 70 million years ago, and the Tertiary Period, about 10 million years ago.[3] So the coal we are burning today represents a process that has taken the Earth between 10 and 70 million years to produce.

There once was a time when there was far too much carbon dioxide in the air to allow the survival of modern-day humans. Modern mammals also require an atmosphere that is rich in oxygen and low in carbon dioxide. The change in atmosphere took place over a period of many millions of years. Plants and trees, which co-existed with early mammalian life as it evolved, removed so much carbon dioxide from the air and added so much oxygen that they ultimately produced the atmospheric levels we enjoy today. Trees and plants therefore generated the perfect mix of atmospheric gases upon which mammalian life, including human life, now depends.

Here is how that was accomplished. Over tens of millions of years, as plants and trees grew and died, the dead wood and foliage fell to the forest floor where it decayed with the help of water. This decaying plant life—and, of course, the carbon within it—was buried in successive layers of peat. Over time, the carbon went into "long-term" storage and was transformed by heat and pressure into coal or oil. It thus took millions of years to remove the carbon from the atmosphere and bury it in the ground in sufficient quantities to produce our present atmospheric balance.

However, humans found the coal and oil hidden in the ground, and discovered that these substances provided a great source of energy. By burning coal and oil, without realizing what we were doing, we have been extracting from the ground the carbon that it took plant life 70 million years to remove from the atmosphere and store safely, and releasing it back into the atmosphere. And we are releasing this carbon at a rate faster than plants can remove it again. We have been innocently and foolishly setting in motion a process that we ultimately won't be able to survive: the uncontrolled release of carbon from safe storage by the burning of fossil fuels. We are reversing a process that took plant life millions of years to accomplish, and we are reversing it so quickly that we are altering the climate of the planet. Warming the climate, while it might encourage some plant growth on a short-term basis, will, in the end, harm the very plants and forests that are a major path by which the Earth can get the atmospheric carbon dioxide out of the air and back into the ground.[4]

One of the ways that we are witnessing the harm to these living forests is in the form of fires—too many fires. While fire has always been a step in the natural part of forest ecology, it typically would occur only once every 100 to 300 years. Today, the fire situation is far more acute.

Major fires have recently disrupted the boreal forest. An area of boreal forest larger than France (56 million hectares) was destroyed between 1980 and 1989.[5] Since 1976, the area burned in Canada has soared to six times the century trend, and to more than nine times the century trend in the alpine and temperate forest of the western U.S.[6] Russian researchers have reported a sharp increase in the area of forest burned since 1985.[7]

In 1989 the worst fires on record scorched western Canada, and the area east of James Bay in Quebec. Fire frequency has increased since 1975 in Alaska, and increased over several decades in Sweden at least until fire statistics were discontinued in 1980. Only in Finland has there been a decrease in the frequency and area of boreal fires.[8]

"The Carbon Bomb: Climate Change and the Fate of the Northern Boreal Forests," Greenpeace International[9]

When trees burn, there is a double loss. Not only is the carbon they hold in their tissues released into the atmosphere, but the living organism that was capable of removing carbon from the atmosphere is killed.

Uncontrolled logging is a problem in the boreal, too. Currently it is estimated that logging in the Siberian forests has reached 15,500 square miles per year (much of it illegal) and fires are burning 4,000 square miles of forest annually. It's been estimated that as much as 10 percent of the carbon we put into the atmosphere is absorbed by the Siberian forests and they are vanishing.[10] Additionally, the boreal fires have put as much as 50 million tons of carbon gases into the atmosphere.[11] One can begin to see how the process of global warming could accelerate very quickly—literally picking up speed— if the processes feeding the trend of increased atmospheric carbon dioxide aren't dealt with very soon.

Humans and other mammals may be able to scurry away from the threat of a forest fire, leaving the trees to burn alone, but ultimately we can no more escape death brought on by a massive failure in the balancing act of the cycle of carbon than a tree in a forest can escape a raging fire. In the same way that

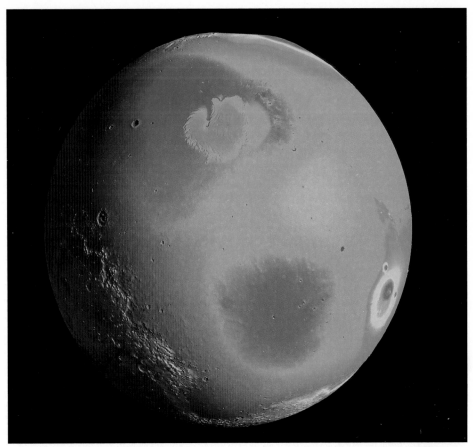

The new Mars Orbital Laser Altimeter images show that the Northern Martian hemisphere holds what was once probably a great ocean covering one-third of Mars. (The blue areas shading into purple represent the lowest areas on the planet, the red and white the highest.) Much of the Northern hemisphere also sits three to five miles deeper into the atmospheric gravity well, meaning that it would have been warmer with higher atmospheric pressures and more protection from meteorites. As such the edges of the Great Martian Ocean are some of the likeliest areas to find evidence of past Martian life. (NASA / Science Photo Library)

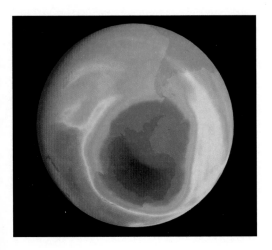

Ozone depletion over Antarctica (1998). Colored satellite map of atmospheric ozone in the southern hemisphere between mid-August and early October 1998. An ozone hole is seen over Antarctica. (NASA / Science Photo Library)

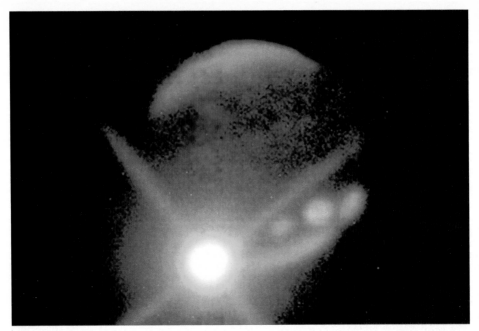

Comet Shoemaker-Levey 9/Jupiter collision. Infrared image showing the fireball created as Fragment K of Comet Shoemaker–Levy 9 hit Jupiter at 10:18 GMT on July 18, 1994. Earlier impact sites are seen to the right of the fireball as glowing spots. (MSSSO, ANU / Science Photo Library)

Chicxulub crater. Artwork of the Chicxulub impact crater on the Yucatan Peninsula, Mexico, soon after its creation. This impact may have caused the extinction of the dinosaurs and 70 percent of all Earth's species 65 million years ago. (D. Van Ravenswaay / Science Photo Library)

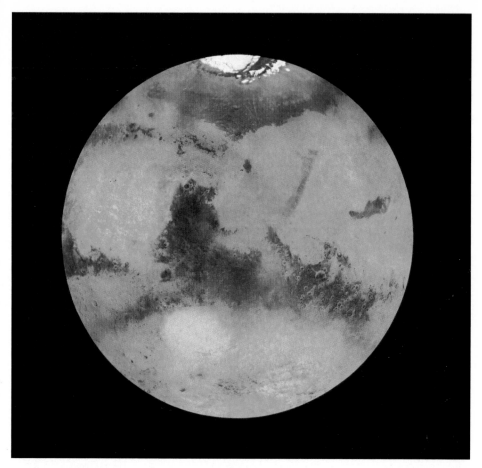

Mosaic of images taken by the Viking I and Viking II probes showing the globe of Mars. (US Geological Survey / Science Photo Library)

Sojourner. Mosaic image of the robotic Sojourner vehicle aboard the Mars Pathfinder. Image taken by the Imager for Mars Pathfinder (IMP) on the day it landed on the surface of Mars on July 4, 1997. (NASA / Science Photo Library)

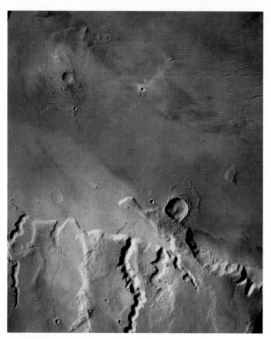

This mosaic of Viking photos shows the boundary scarp between Mars' ancient cratered highlands and the northern plains to the east of the Mangalla Valles. Ancient river channels run north through the highlands. The latest NASA information indicates that the less cratered northern regions may have been the site of a paleo-ocean on Mars. (US Geological Survey / Science Photo Library)

Forest fires, Brazil. The Amazon, Roraima State. The savannah area north of Boa Vista. (John Maier / Still Pictures)

Erosion, Brazil. When the forest is destroyed, the top soil rapidly deteriorates. (Mark Edwards / Still Pictures)

Surface of Mars. Mars Pathfinder mosaic image of the dusty surface of the planet Mars. The area around Pathfinder features rocks such as the ones named "Flat Top" (at upper right) and "Wedge" (at upper left). Image taken by IMP on the third Martian day (Sol 3) it had spent on Mars since landing on July 4, 1997. (NASA / Science Photo Library)

Cameroon. Aerial view of the Borombimbo crater lake close to Kumba in the South West Province. This is one of a line of crater lakes through the rainforest of Cameroon, one of which (Lake Nyos) released a deadly cloud of carbon dioxide killing 1700 people in 1986. (Edward Parker / Still Pictures)

Forest fires, Indonesia. Sumatra, Padang, Tanjung Bungus. Kerinci passenger boat stranded because of the thick haze. (Tantyo Bangun / Still Pictures)

Solar Power Station. Dish-shaped solar power reflectors at a solar power station at White Cliffs, Australia. Each dish focuses radiation from the sun onto a thermoelectric generator positioned in the focal point. A computer steers the dishes to ensure they face the sun throughout the day. (John Mead / Science Photo Library)

Fusion Research. Electrical discharges over the Particle Beam Fusion Accelerator (PBFA-II) during one of the accelerator's "shots." The target is a gold cone, at the focus of which is placed a small pellet of deuterium and tritium. The beam causes the target to emit very intense X-rays which collapse and ignite the pellet, causing a nuclear fusion reaction. (Sandia National Laboratories / Science Photo Library)

Forest Fires, Indonesia. Sumatra, near Bukit Tigapuluh. Kubu tribesman surveying the burning jungle. This area of forest has been used by generations for hunting and gathering medicinal plants. (Mark Edwards / Still Pictures)

Temperate Forest, France. Trees are the ultimate representation of life on earth. Through the planting and the preservation of trees, Garden Earth will be reborn. (C. Comte / Still Pictures)

a tree can't go anywhere to get away, neither can we. There is nowhere to go when our planetary system fails. Not even Mars.

In the blue sky of morning, the jet fighters of the elite New Mexico Air National Guard were roaring off the runways over the buildings of the Sandia Laboratories. They flew their planes like arcade gamers and we half expected one of them to crash into our parking lot someday. The roar of the departing fighters echoed in the hallway of the main experimental building, or "hangar" as it was more accurately known. I was looking for John McDonnell, or "John Mack" as he was affectionately known at Sandia. John was a trained geologist, who had somehow ended up at Sandia as an engineer on the Particle Beam Fusion Apparatus (PBFA), the enormous device that Sandia was using to try to achieve nuclear fusion. They were getting ready for a "shot" and I knew that John Mack would be totally involved. I stood in the hallway as more and more personnel filed out of the main experimental bay, milling about in subdued excitement. Inside the main bay was the enormous fusion device consisting of three concentric rings, the outermost 100 feet across.

The outer ring was the ring of oil, in which huge banks of capacitors (storage devices for electric charge), each as big as a 10-gallon gasoline can, were bundled together into structures bigger than household refrigerators and charged to 50,000 volts each. These were called Marx banks, after their inventor. The voltage was so high that the capacitor bundles had to be submerged in vats of golden oil 15 feet deep to prevent arcing. Each Marx bank stored as much energy as several sticks of dynamite and, when triggered, would release its stored energy in a few billionths of a second. Each of the 20 Marx banks would fire their energy inward, to the middle ring, a ring of water.

Here, in crystal-clear, purified water 10 feet deep, sat tubes and spheres of aluminum large enough for a man to stand in. In fact men swam inside them, to make inspections and to clear bubbles of air that could disrupt the flow of lightning-like electric power that would course through the aluminum after the Marx banks fired. The divers were a rugged group, with the classic build of swimmers. In winter, the water stored in tanks outside the experimental hangar would plunge in temperature to near freezing, but still the divers endured. Wearing wet suits and breathing through air hoses, they went down into the cold water before and after each shot, checking electrodes and sparkgaps, and chasing bubbles.

The ring of water received and held the electric charge released by the Marx banks for a few billionths of a second. The water polarized around the ring on its aluminum electrodes, and for a brief instant, the entire ring was a perfect high voltage insulator. The massive bolts of electricity were tightened and concentrated in the ring of water before passing into the third ring.

This innermost and tightest ring was a ring of steel and vacuum where the bolts of electricity were directed down double concentric pipes—like spokes of a vast wheel. The electricity rode into the middle of this ring as a thin film of lightning on the inner pipe. The voltage was now so high— literally millions of volts—that no material substance could insulate it, and electrons could form spontaneously out of the vacuum. What guided the electricity on the inner pipe and prevented it from leaping to the outer pipe was the crushing pressure of its own magnetism that wrapped around it like rolled cable.

This use of magnetic containment is a wonderful example of Russian genius fueling American scientific progress. The pulsed power engineers at Sandia had made lavish and ingenious use of the concept of "magnetic insulation," which was invented by the Russians to concentrate and focus bolts of electrical energy. The idea was that when electricity was focused tightly enough, its own magnetic lines of force would enclose it like a finely spun fabric and prevent any electrons from escaping.

All of this elaborate, massive engineering was created so the 20 Marx banks could fire their inwardly directed bolts of electrical energy, and so those bolts would focus inward to tighter and tighter concentration, and finally meet in the core of the innermost ring of the PBFA. There, like rays of sunlight brought to focus by a lens, the bolts of energy converged on one point, passing inward from the ring of oil to the ring of water, to the ring of vacuum. Imagine the energy of a ton of dynamite, first focused at a 50-foot radius, then 20, then 10, and finally compressed into inches, then to collide in a cyclonic implosion whose density of energy rivaled the center of an exploding star. In that center collision point was a small pellet of fusion fuel, that would, if all went well, be both super-heated and super-compressed to release, in a single, tiny fraction of a second, vast amounts of star-like energy in a single burst, whose sum of fusion energy would exceed even that of the total of all the electrical energy focused on it.

In the hallway outside the hangar we stood or sat with our backs against the wall, voices hushed, waiting. John Mack came bounding out of the now darkened doorway, the last man out. He grabbed an intercom microphone from the wall and yelled, "The bay is clear!" The lights dimmed in the hallway as the door to the main bay was bolted and flashing red lights suddenly flared into brilliance near the ceiling. "Twenty thousand . . . thirty thousand . . . forty thousand . . ." a voice announced as the entire bank's electrical voltage rose. Everyone tensed, and covering their ears braced themselves against the walls. I cupped my hands over my ears too and leaned against the wall for support. "Fifty thousand volts, arming . . . five, four, three, two, one, ZERO!"

A tremendous clapping crash filled the entire building and the concrete floor jumped. The hairline gaps around the door to the main bay filled for an instant with blinding blue-white light. In mere seconds the door was flung wide open and in rushed the stalwart John Mack, his face a mask of determination. He was always the first man in after a shot.

In the hall there was a wave of happy relief. "Sounded good. Nice and sharp," said a diver. "Yeah, not one of those rumbles like when it misfires," said another. I muttered a prayer of thanks. I wanted badly for the project to succeed, despite the fact that I was not yet directly involved. The PBFA was part of the nation's effort to harness controlled fusion, which most people agreed would be the best form of nuclear power. But it was hard to do.

After John Mack had given the all clear, the workers trooped back in through the door. He emerged from the bay smelling of ozone and smiled at me as I waved to him. I had asked him to review some articles of Mars science for me since I was not a geologist and he was. He sat down beside me on a bench in the hall and gave me his analysis. "I read the two articles you gave me. Your so called 'face' is in an area called Cydonia on Mars." (That was the first time I had heard the word "Cydonia.") "I have an atlas of Mars at home which shows that it's a region full of mesas, where the land changes from highlands to smooth plains," he continued. "Your article on the soil analysis by the Viking lander was really interesting. The soil composition matched red montmorillonite clay. That's ocean-bottom clay."

"Really? Why is it red?" I asked, suddenly realizing that I was asking the age-old question, Why is Mars red? (At a distance, Mars does have a slight reddish cast when compared to other planets.)

"It's red because of oxidized iron—iron in its ferric or highest state of oxidization," John Mack continued. "Iron in the ferrous state is black. Black ferrous iron is like black lava rock. Then it weathers to clay and reddens because the oxygen in the air oxidizes it. That's why the desert is red here, around Albuquerque."

"So you mean Mars could have had oxygen in its atmosphere in the past, like Earth?" I speculated.

John Mack looked puzzled, as if it was a trick question. He was a cautious character and it was sensible to proceed carefully. "It's possible," he said grinning. As he said this, a senior staff member approached to ask John Mack a technical question, so I patted him on the shoulder in thanks and walked away. However, as I walked, I mulled over his words carefully. Armed with this new, simple understanding I could now see the Mars surface, even the exposed layers in the Vallis Marinaris, in a new light. These places were bright red from the highly oxidized iron in the soil, I thought— and although it was probably just ozone that still lingered in the air from the fusion shot, I thought I caught a whiff of oxygen, too.

In St. Louis, Christy Booker had been given something new to help her breathe. A new way of taking medication was easing her breathing without making her constantly agitated. Medical science had given Christy a new inhalation device, developed to help people with asthma deliver medication directly to their bronchial tubes. Primitive by today's standards but a great advancement over the adrenaline the doctor had administered or the Theophylline she'd ingested with its continuous plague of side effects, inhalation devices significantly improved the quality of life for children with asthma.

To use the device, a powder-filled gelatin capsule was broken inside a special chamber which released a fine powder. With a puff of air, the loosened powder was then swiftly inhaled directly through the mouth. The taste of the powder was very bitter but it quickly improved Christy's breathing most of the time. The overstimulation that was a side effect of the medication was very short term, lasting only 20 minutes or so.

While the introduction of this device was a revolutionary advancement for Christy, a disturbing trend was appearing in the world at large.

The incidence of asthma in children was rising. Hospital emergency room visits for asthma were skyrocketing and more children were suffering and dying, particularly those in urban areas. No one knew exactly why, but it was fairly certain that air quality had something to do with it. The rise in asthma rates continues to this day.

The increasing number of children with asthma is a trend that we are very quick to distance ourselves from, even though most of us know someone who has difficulty breathing. We distance ourselves by pointing out that genetics may play a role (it can be a factor), or by saying that these children have allergies (many do, often multiple allergies), or by pointing out that asthma affects a higher percentage of those children who live in poverty, or children who live in the inner cities, or those who are African American (all of which claims have some level of validity). Nevertheless, we should not miss the point. Worldwide air quality is also a factor, perhaps the major factor. While, perhaps, for a variety of reasons, these children are more sensitive than we are, like the canaries in a miner's cage they may be telling us that there is something very bad in the air. And as much as we may want to, we cannot distance our lungs from the fact that we are breathing the same air. If we were in a mine, we'd get out if we could, and we'd take our children with us. But there's effectively nowhere to run to when you're already on the surface, breathing from the only ocean of air there is.

W e couldn't have been more excited if we'd actually been going to Mars. Vince DiPietro and I met at the gate at the airport in Denver and hopped into a rented vehicle. Though it was only a car, we felt like we were jetting off to present our scientific paper on Cydonia at the Case for Mars meeting in Boulder, Colorado. The paper I'd drafted had attempted to summarize the work of the Independent Mars Investigation Team (IMIT), and it had been accepted for presentation after peer review. Although I'd presented scientific papers at conferences before, I was far more excited about this one. Between the sheer thrill of doing Mars science, particularly with such fascinating findings, and actually meeting Vince in person for the first time, the conference in Boulder was a turning point in my life.

Writing a paper expressing IMIT's findings had been challenging. The group had two different, though not necessarily conflicting, ways of looking at the various features at Cydonia. One group was interested in focusing on alignments, looking at how the features related to each other or to other celestial bodies, much as we do with archeological sites such as Stonehenge. I myself supported another line of inquiry. I wanted to know if, given the past climate of Mars, indigenous intelligent life could have been possible there. Was there ever a time—a long enough time—when this could have happened? However, others in our group preferred not to limit our thinking to indigenous life. Therefore we had to compromise and write the abstract loosely so as to both present our overall case and not exclude anyone's point of view.[12]

This Case for Mars conference, in July of 1984, the second to be held, was the brainchild of the Mars underground, a group of students and space professionals who wanted the U.S. to send humans to Mars. The basic thesis was that, as the result of an explosion of knowledge about Mars from the Viking probes, the red planet apparently held everything necessary to support human life. Mars was not hostile to humans. With ingenuity, Mars could provide everything humans need: air, water, shelter, and food—but this possibility would mean something only if humanity recognized that going there would serve some greater purpose. The results of the Viking probes, both orbiters and landers, were encouraging. The North Polar cap appeared to be several miles thick and made almost entirely of water ice, enough water to cover the planet's surface to the depth of several yards. Together with the carbon dioxide and nitrogen in the thin atmosphere, the water could be used to grow crops in specially constructed shelters. Mars appeared to contain every sort of mineral deposit that the Earth contained. Therefore, if equipped with a source of energy to supply power, a human colony, resembling those at Antarctica, could be established and prosper. It was a bold and enthralling goal.

There was a wide variety of scientists, engineers, and philosophers attending, all trying to lay the groundwork for this vast futuristic undertaking, and we were right in the middle of this creative foment. Cydonia could be seen as part of a larger effort to expand the human world, to break out of the deadlock of the Cold War by changing the problem. Cydonia might offer a compelling reason to send humans to Mars, one that could hasten a Mars landing by decades. So although the organizers of the conference took a gamble when they allowed us to give a presentation, there were also significant reasons to include us.

We marched into the exhibit hall as a team[13]—ready to assert our findings. These were to be presented as a poster paper displayed on a large bulletin board in a room with many other papers. This may seem a somewhat humble way to present, yet if the paper attracts a large number of interested people, especially for the summary talk, it can be the most intense and useful of all the means for disseminating ideas.

Due to Vince's extraordinary skills as an organizer, our paper was displayed on our bulletin board early, and it was a wonderfully skillful presentation.[14] The whole story was covered. We displayed images of the surface of Mars, showing evidence of past water—including a probable old ocean. There were also close-ups of the "face" and a variety of enhancements, including images of the "pyramid" and "fortress," the colloquial names given to two of the other unusual objects on Mars.

A large crowd quickly gathered at our post, including two senior scientists. One of them, a rumpled looking character with long brown hair, was beaming ear to ear at our poster; the other, a tall patrician looking gentleman with a huge mane of silver hair, focused on me with a glowering hostile stare. I asked who the silver haired character was and was told that he was a NASA official, the "chief enforcer" for the NASA space station project. Intimidation seemed to be the tactic of choice. However, since I was granted the honor of being the chief presenter and was fortunate to have the obvious interest of most of the crowd, I proceeded to give a summary talk to the group as if the NASA official were the only one there, to drive home the point that we weren't easily intimidated.

But that wasn't all that was brewing during my presentation. The person giving the talk at the poster next to us appeared to be very annoyed with us, possibly because we were attracting a larger crowd. It was as if he were trying to shout over me, trying to make it difficult for me to be heard. I turned to Vince after my talk and asked, "Who is that prickly SOB?" Vince replied, "Some character named Malin who has a fancy Mars camera design." If I had known what events would later unfold, I would have made a more conscious attempt to annoy him, for our paths were fated not just to cross, but to collide.[15]

As we discussed the paper with interested individuals, we found that we had indeed attracted an unusual level of interest since crowds stayed even into the lunch hour, which we celebrated as the ultimate acknowledgment. One bold character, a scientist named Steve Squyres, engaged me in a sort of impromptu debate in front of the poster. He was not hostile. He was skeptically curious, which I regard as the most useful of all attitudes for a scientist. For a scientist to be curious but not skeptical is ill-advised, but to be skeptical without curiosity is to be merely a scoffer and to deny the essence of science. Steve Squyres told me that Carl Sagan and another fellow named Fox had already searched Mars for signs of civilization using the Mariner 9 images and found none, and that they'd published their results.[16] I was immediately intrigued. This meant what we had done was merely following in their footsteps. In other words, the idea of looking at Mars images to find signs of a dead civilization was already considered a legitimate exercise. "You should talk to Carl," he suggested.[17]

I placed a call to Carl Sagan's office at Cornell University to locate the references to his search for signs of civilization on Mars. I intended to write a scientific article based on our Boulder paper, and it was both scholarly and politic to include Sagan's work. To my great surprise, I found myself talking with Carl himself.

He was extremely polite and well spoken, and basically asked me why I was stirring up this trouble over Cydonia. I told him we had investigated and found objects quite suggestive of archeology. Then I asked him why he had bothered to look at images of Mars for signs of archeology if it was such a foolish thing to do. He chuckled and said that he did it because JPL would have looked pretty foolish if someone else had found such a civilization later, and they'd overlooked it. We then discussed the riddle of Mars' past climate and the problems of determining age on Mars by counting craters. It was a great honor to engage with such a person, and he raised many good points which have been incorporated into our later works.

Shortly after the Boulder conference, astronomer Hal Masursky, a member of NASA's Viking mission team, came to give a lecture at Sandia Labs which I attended. Remembering me from Boulder, I was invited to join him for lunch, during which he remarked with a smile, "We didn't know what to think about your pictures, but we really liked your ocean." He suggested that I publish the Mars ocean hypothesis, a recommendation I followed, encouraged by his support.[18] At that moment I realized that I had landed on another planet, scientifically speaking, and given the circumstances, perhaps just as remarkable, I'd been able to walk away from the landing.

CHAPTER 6

Virtual Unreality

The dream of Pharaoh, lord of all Egypt, was terrifying. He saw seven cattle, strong and fat, devoured by seven emaciated cattle, who grew no fatter. The dream was repeated with seven ears of wheat, full and heavy, and seven ears, thin and parched, which devoured the fat ones, but grew no fatter themselves. Together the cattle and wheat symbolized the staff of life and wealth of Egypt, perhaps the greatest and best-governed nation in ancient times.

Fortunately for Egypt and the then known world, Pharaoh got excellent advice. Joseph, the Israelite, interpreted the dream as a warning from God that seven bountiful years were coming, to be followed by seven years of famine. Joseph advised him to build a system of granaries in which to store a portion of the grain in the years of bounty, so Egypt could survive the seven years of famine. Pharaoh followed Joseph's advice, and the people of Egypt and the surrounding region, including Joseph's relatives, were saved from hunger, for the famine was widespread and very severe.

This simple story from the earliest portion of recorded history illustrates the actions of a responsible government, and the importance of an unspoken character in this story: the cat. The Egyptians' main source of food was bread that had been leavened by allowing yeast to release carbon dioxide into the dough to make bubbles and cause the dough to rise. The Egyptians also loved beer, another foodstuff made from grain. The beer's grain mash was fermented, allowing yeast to release the carbon dioxide bubbles which produced natural carbonation. Since bread and beer were essentials to their diet, the Egyptians had to store large quantities of grain, but without cats they would never have been able to preserve it for seven years.

Egypt has a dry climate, so if grain is stored in stone or clay vessels or in an enclosure with a stone floor, it will not get moldy. In fact, it is this very dryness, combined with even a minor climate change, which may endanger the Nile in the future, but that certainly wasn't an issue in Pharaoh's day. Nevertheless, how like Pharaoh's dream—to have carbon dioxide, like that released in bread and beer, return to devour the great river.

Egypt, like most places, had mice and insects, and in order to protect the grain stores, it was essential to have a strategy for coping with these persistent pests, who were nearly impossible to keep out of the granaries. This is probably why the cat, a somewhat socially aloof creature but a very effective small predator, was domesticated in Egypt. Cats, who preferred insects and mice, provided precisely the solution the Egyptians needed.

So cats and humans formed a symbiotic relationship that benefited both species and the little felines became very popular in Egypt. They were honored, admired and no doubt loved, and cats have lived closely with humans ever since.

And so have rats of all kinds . . .

It was a bit unusual for a major financial organization. Nevertheless, the Securities Exchange Company, which was taking in over a million dollars of business a week at the time, was operating out of the Bell in Hand, a bar-room tucked into Pi Alley just off Washington Street in Boston. Regular customers didn't seem to have any problem finding the company at the new address and were willing to overlook the unusual location as long as they got their money.[1]

In fact, numerous people, including numbers of very tough-minded, financially adroit Bostonians, had managed to remain oblivious to some of the more unorthodox aspects of the Securities Exchange Company for months. For example, although the company owned 38 percent of the

Hanover Trust Bank, which was obviously a major asset, the entire organization was administered by 18-year-old Lucy Mell.

Although Lucy may have been a very talented young woman, it is difficult to imagine that a bookkeeping clerk just out of high school had the kind of personal authority required to manage a staff of 30, working in three offices in Boston, with branch operations in several other cities. Nonetheless, when you are making money hand over fist, you tend to overlook some of the rough edges. Who wants to argue with success?

The nature of the business—investing in postal reply coupons—was a simple strategy based on differences from country to country in the value of pre-paid postage—like an early form of investment in foreign currencies. Coupons could be purchased in another country for a cent or two and redeemed by the US Postal Service for the face value of the coupon, often three or four times the purchase amount. Securities Exchange Company was willing to share these wonderfully high levels of profit with their customers. The most important thing to understand about the business isn't what they did but that investors were making lots of money—$500 in interest on a $1,000 investment over 90 days was what their notes promised, but the company routinely paid back the principal with the full 50 percent interest in only 45 days.

As you might imagine, it was as extraordinary then as it is now for a company to give customers more than they promised, especially since the interest rates were already unbelievably high. Satisfied investors, glowing with the enthusiasm of money-making success, quickly reinvested their profits in new notes and helped to make Securities Exchange Company the talk of the East Coast over the summer of 1920.

However, the company came under increased scrutiny as a result of an investigation by the Massachusetts District Attorney. Lacking material evidence of wrongdoing, the District Attorney wasn't able to do much other than demand an audit, since not a cent had been lost to a single investor over the entire history of the company, and many had achieved great wealth. The company's owner argued mightily that the government had no business interfering in his relationship with his customers, and those who were busy raking in the profits undoubtedly agreed. Frankly, if there was any crime involved, it appeared to be the crime of being too successful and, as a result, antagonizing those who would stifle free enterprise. However, the owner of the Securities Exchange Company did consent to temporarily halt the acceptance of new investments, and, when word of the mandated audit got out in the press, it created a bit of a panic.

The following day a flying wedge of creditors jammed through the door of the School Street offices of the Securities Exchange Company and it took seven officers of the law to re-establish order. Even with the confusion, the company promised to pay all interest and principal on mature notes and make refunds of the full principal to anyone who wanted it, whether the notes had matured or not. It seemed a generous offer, particularly since there was no legal obligation to make refunds. Further, so many customers had shown up that Monday, July 26, that the following day, to accommodate more traffic and in an effort to keep order and avoid a similar near-riot in the crowded office, the "branch office" was opened in the Bell in Hand in the alley directly behind School Street. That's why customers were lined up early outside the door of the Bell in Hand, forming a queue that snaked down the alley all day long.

Once inside the bar, clerks reviewed a customer's paperwork and those approved were paid at a cashier's cage that had been freshly constructed just inside the backdoor. Everything went smoothly. There was little discussion and few disputes. Over a thousand transactions were enacted that day and while some were clearly in line to get refunds, others were there to collect on maturing notes, with many of those oh-so-satisfied customers reinvesting the profits.

The following day the line rematerialized. This time, however, the Security Exchange Company served coffee and hot dogs to the thousands who were waiting. Reassured by the appearance of "business as usual," even under difficult circumstances, many reassessed their decision to withdraw their funds and went home. Others, eager to take advantage of the situation, wandered through the crowd offering to purchase notes for a premium from those who wanted refunds. Over the course of the day, a million dollars was paid out and an impromptu press conference defending the business inspired new confidence. Angered by the injustice that was being perpetrated by the government, an ad hoc alliance of customers who had profited from the Securities Exchange Company held a meeting and drafted resolutions supporting the company.

By the end of the week there were only a few customers in line waiting for the offices to open. The run had finally been thwarted and it looked like the Securities Exchange Company had weathered the storm. The world was once again safe for private enterprise . . . but not for long. By the following Monday another major run was trigged by another negative newspaper story. The company once again scrambled and managed to pay off all customers, paying out a fortune in the process, but miraculously, by the end of the week the run had yet again subsided and it looked like the problems were over once and for all—except for a small matter concerning a $441,878.07 overdraft. And the auditors' discovery that the whole operation had been a scam from the very beginning.

Charles Ponzi had operated the Securities Exchange Company based on a very simple strategy. Original investors were paid interest from the principal that new investors were paying in. As long as there were always new investors, the old ones were paid off, and they then provided the positive word of mouth that attracted further new investors. Also, more often than not, they would reinvest their newly made profits, confident that they could make even more, which of course put the money back in the ever-increasing pool. Essentially, a "Ponzi Scheme," as this type of con eventually became known, requires reaching further and further for assets to cover its inherently bankrupt nature, but a talented "financial genius" like the charming ex-convict Charles Ponzi was able to keep it going for a long time. Unlike many others who have used this scheme, Ponzi didn't abscond with the funds. He stayed to the bitter end, and went to jail protesting his innocence. It was the ultimate con, one which seduced even the con artist himself. Ponzi seemed to believe his own propaganda about the grand, glorious intentions of his business, despite the fact that rarely, if ever, was a postal coupon ever purchased or redeemed.

Then again, the Securities Exchange Company didn't need postal coupons when it could operate on greed. Unfortunately, when these types of schemes go belly up, which they inevitably do at some point, it is always the most recent investors that are left penniless. The theft doesn't actually happen until the very end. Until then everything can look fine. In Ponzi's case, it looked fabulous. This is the important thing to remember about a Ponzi scheme. Everything works great up until the exact second when it suddenly doesn't. Then, instantly, everything collapses and the ones at the end of the game are left wondering what befell them, while the investors who have already been paid off are safely at home, thinking about how well they did, unwilling to share in any responsibility.

It is vital to understand how a Ponzi scheme works, because we are all part of one that's on the verge of collapse, and most of us don't even have a clue about what's ahead, yet. We are part of a scheme that has been going on for over a hundred years, longer than any of us have been alive, but we are very near the end.[2] This Ponzi scheme is so much a part of our daily life that it just seems normal to us. Because it is as invisible as the air, many of us would protest that there isn't really anything to be concerned about. Regretfully, those who protest will only have to wait to see the miscalculation.

If we were investors in the Securities Exchange Company, we would be just at the point when the very first negative articles were appearing in the newspapers, only in our case the article would be saying something like "scientists discover indications of global warming," or "record damage caused by weather," instead of "Massachusetts District Attorney investigates reports of financial irregularities." Just like the investors who made lots of money in Ponzi's scheme, there are many of us who will look at all the evidence we can see and conclude that there isn't a problem. Everybody is making money. The U.S. Stock Exchange is at record highs. Everything is just fine. In Chicago, we didn't even have to shovel snow last winter.

Ironically, that's how it always looks with a Ponzi scheme—until it doesn't. Suddenly everything falls apart. One minute you are on your way to millions, then, at some critical moment, something fails—the capacity of the carbon sinks, the crops, the ocean currents, the rain. Nature's line of credit dries up and our most recent check, written on a long overdrawn account, is returned. The life we have always known quickly collapses.

In a suit and tie, I cross the street near the White House in Washington, D.C. It is a sparkling, clear morning in July, 1985. I've moved to a new city and have a new job, but my work is much the same. I'd managed to make a remarkable escape from Albuquerque, selling my house and moving to the suburbs of northern Virginia, due south of the Pentagon. The Reagan arms buildup is in its sixth year, and in the air is a mounting sense of victory. I am contributing to that victory. However, there is something else in the air. The lost cry of a ghostly cat echoes on the breeze . . .

The cat was born in a thought experiment that was posited by an Austrian physicist, Erwin Schrödinger, who was awarded the Nobel Prize for physics for 1933. The experiment was designed to illustrate the fundamentally paradoxical nature of quantum mechanics, which describes the world of sub-atomic particles.[3]

In this experiment, the cat is placed inside a steel chamber with a fragile bottle filled with a deadly poison. Poised over the bottle is a hammer connected to a mechanism which is triggered by the decay of a radioactive atom—an event that may or may not occur at any time.

Although we can't see inside the chamber, it's obvious that there are at least a couple of possibilities for the cat after an hour or so. It will be either alive or dead, depending on whether or not the atom has decayed. However, in the peculiar domain of quantum mechanics, since we don't know (and can't measure exactly) what state the radioactive atom is in, it could also be in a third state—a "superposition" of each of the other states (decayed/undecayed). So as long we don't open the chamber, we have to assume the atom could be in a superposition. The question then arises: Is the cat, like the atom, in a mixed alive/dead state? What has happened to the cat?

The probability field is thus of the virtual cat alive and superposed, like a double image, on the virtual cat which is incomprehensibly dead at the same time. Only when we open the chamber/coffin do we know if the cat has actually been killed. At this moment of knowledge or measurement, the state of the cat becomes singular and precise, but until that moment we can only calculate probability. There is the field of the cat dead, and the field of the cat alive, and, at the time Schrödinger developed this concept, in 1935, these fields were all that were proposed. It is a clean and tidy construct, this doomed cat of Schrödinger's, both dead and alive, and it has been the subject of many an erudite discussion.

Some have suggested that this doomed cat in the chamber with the poison vial of gas ready to be shattered upon the decay of a single radioactive atom, was the product of a mind steeped in scholarly detachment, but this was not so.

At the University of Berlin, beginning in 1932, mobs of Nazi youths went around beating up anyone they believed to be a Jew. The police did not intervene but, instead, erected police lines around the campus, placing it seemingly under siege. Passes were issued to faculty members so they could pass through these lines—a daily reminder of the Nazis' growing distrust for academics. Schrödinger was a man who, on occasion, had intervened to help Jews under attack, but as he saw the machinations of the Third Reich closing in around him, he prepared to flee Berlin for his life. Although he was not Jewish, he saw that the life he had lived was already doomed.

Schrödinger knew that the death of his quantum cat would not be clean and tidy. The virtual dead cat was already howling its death cry and hurling itself at the walls of its tomb as the poison gas filled its lungs. It clawed the walls frantically in a last burst of strength. As the glass vial shattered, the cat suddenly gained knowledge denied everyone else. Its great burst of insight and strength came too late, however; its death was neither quick nor painless, but merely efficient. So, in 1935, the cat's ghost was already screaming in the halls of the University of Berlin where Schrödinger was working when he conceived this experiment, two years before he fled in horror and terror from Germany to England. He had been given (and taken) warning from the ghost of a cat screaming the dangers of a poison gas. Could Schrödinger, in creating his thought experiment, have detected the evolving probability field of the holocaust?

It was a beautiful July morning, not in Berlin but in Washington, D.C., and the virtual cries of Schrödinger's cat were somehow present. It had prowled the halls the previous night. It was a massive cat, mutated as big as a bear, as strong as a rabid wolf, burned beyond recognition and so charged with radiation that the air glowed blue around it from Cerenkov light. We had progressed beyond poison gas vials in 1985. Our virtual death would not be from the decay of a single isotope, but from the annihilating implosion of several pounds of plutonium. I was working on the Strategic Defense Initiative (SDI) then, in the attempt to avert that implosion. SDI was a great exploration of the hairline edge of thermonuclear holocaust. As a consequence, my work was nothing if not challenging.

I was ignoring the ghost of the cat that day. I had other problems, I was lost. I gazed down the street in both directions and looked for the most splendid building I could find, figuring that would be the National Academy of Sciences where the Case for Mars conference was to be held. However, I was surrounded by splendid buildings. I guessed at one and crossed another street to get to it. It was built of beautiful white marble but there was no sign to indicate what the building contained The only human being in sight was a stocky man in a green suit standing on the corner. I approached and asked if the building behind him was the Academy. My question startled him, as if he had been daydreaming and had not noticed my approach. He jerked his head around and looked at me with wide eyes and stammered "yes" before turning and walking away.

I was pleased that I was already at my destination, so I entered the stately front doors. Only when I took a seat in the auditorium for the opening presentation and the man in the green suit mounted the stage did I realize that I had asked a Russian cosmonaut, Valery Kubasov, for directions in downtown Washington, D.C.

Inside the conference, I joined Vince DiPietro and a friend and we planned which talks we wanted to attend. It was a great day for me to explore Mars. I needed to escape. My new job was going well, but my marriage of 12 years had been added to the list of casualties resulting from my investigation of Mars. At home my wife was packing to move back to Albuquerque without me, and my four-year-old daughter had approached me the previous night, with tears in her eyes, and told me "Daddy, I like Mommy." At this, I tearfully embraced her and told her, "Yes, of course you do. Of course you do." I knew that the choices I had made in life had painful consequences for me and for those I loved, but the pursuit of science is fundamental to who I am.

To be detached from life is one of the great privileges and curses of the scientist's calling. To sit in one's bathtub like Archimedes and notice the water spill over the sides and have a full-blown understanding of what you have witnessed hit you in a blinding flash—that is the joy of a scientist. It was that joy that launched Archimedes out of the tub, to run naked and dripping-wet down the street shouting "Eureka! Eureka!" (I have found it!) all the way home from the public baths. That his joyful discovery led immediately to the exact determination of the density of the metal in the King of Syracuse's crown, and thus to the execution of the royal jeweler for theft, is the curse of the scientist's craft. The heavenly insights sometimes lead to horrible consequences on the dimension of mundane reality. It is said that Archimedes lost all taste for practical problems after that and focused instead on pure mathematics—an abstract domain where the realities of social consequences were unlikely to intrude. It is truly the hope of scientists that the consequences of our insights will be good, that we will discover penicillin, that we will invent the microchip, that we will advance the fortunes and future of humanity.

The study of Mars demanded and delivered both passion and detachment in equal measure. The conference was a thrilling place, with talk after talk spinning the shining visions of human missions to Mars. While sending humans to Mars will not directly solve problems on Earth, we seem to know in our bones that human evolution is inexorably entwined with the drive to explore space. Mars is the most obvious destination.

At the conference it was posited by Carl Sagan that if Americans and Russians created a joint mission to Mars, then they would have one more good reason not to destroy each other. Prior to this, Carl had been a staunch opponent of human spaceflight as a logical way to explore the cosmos, preferring robotics. But human events had altered his opinion. Perhaps his change of heart was not logical, but neither was the Cold War. In fact, it was reckoned by many at the conference, including me, that making a joint U.S.–Soviet mission to Mars might be the most sensible undertaking both countries had pursued in a long time.

I encountered Carl out in the main hall after his talk and congratulated him. He greeted me cordially. We'd carried on a lively, if not always happy, correspondence since we'd first met. At one point, I sent him a rough draft of an article on Cydonia, only to have him rip it to shreds and remind me of the dismissive statement of Wolfgang Pauli, an influential quantum physicist who taught in several German universities early in this century. Pauli, Carl recalled, had said to the colleague with whom he was arguing, "You are not even wrong!" Carl later apologized for his reaction, which I appreciated. Even the most rigorous of scientific minds may change their opinions over time, as Carl had done over manned spaceflight. Ironically, Wolfgang Pauli, known for his take-no-prisoners rigor in science, eventually changed his as well and became something of a mystic toward the end of his lifetime.

However, both Carl Sagan and I were intently focused on things about Mars that could be quantified. The questions about Cydonia had become questions about Mars' history. The argument became one of how to interpret the Viking Mars data. Paramount was the question of how long liquid water had flowed on Mars' surface. Was the history of Mars lunar, with a liquid water era lasting only for an early fraction of its history? Or had Mars been primarily terrestrial, with liquid water, and, thus, conditions conducive to life as we know it, over most of its history?

Carl eagerly agreed that Mars had begun Earth-like. We even agreed that during this early period life had probably begun on Mars in much the same way as it had on Earth, as evidenced by the bacterial "microfossils" in the most ancient of Earth's rocks. But for how long had Earth and Mars evolved along a common track? For how long had Mars been Earth-like? We both knew that this was the crucial question about Mars.

Everything on Mars, including dried-up water channels, was age-dated by counting how many craters could be seen per unit of area on its surface. As a general calculation, lots of craters indicated a geologically ancient water channel, few craters implied a geologically recent channel. So we were ultimately arguing about one number, the cratering rate on Mars.

The cratering rate on Mars was assumed to be close to lunar; however, if the average rate of Martian cratering was in reality half or double the assumed rates, it would make a difference of billions of years in the estimated age of certain Martian features. We were arguing about the rate of crater formation on Mars, but this was something that could not be known for sure, only estimated.

The cratering rate on Mars, the crucial number that determined whether Martian history was Earth-like or Moon-like, was estimated based on our Moon. The Moon, like Mars, is covered with craters, and someone has counted them all, at least in the most important areas. In fact, the Moon, per unit of area, has many more craters than Mars because there has never been water erosion there—ever. The Moon and its nearest neighbor, our Earth (with almost no craters), represent the two extremes of history. However, it wasn't enough just to count craters. The raw number alone meant nothing. Fortunately, humans have been to the Moon and brought back rocks that could be dated by the weak natural radioactivity found in them. The rocks from known areas could be age dated, and with the crater counts of those areas also being known, an estimate could be made of the rate at which craters formed on the Moon. This meant that, at least at one point in the solar system, we knew how fast craters of various sizes formed. So if you saw a surface with a certain number of craters per unit area, you could get a good estimate of its age.

There were, as usual, complications with this scheme, and many important details to be considered, but the Moon gave us a standard cratering rate in the solar system—the only one we had. This method was immediately applied to Mars and, voilà!, Mars gave lunar ages—the Lunar Mars model was born. There was, however, a slight problem with this scheme that even Carl acknowledged: Mars was not located next to the Moon. Mars was much further out, right there next to the asteroid belt.

Another interesting luminary I met at the conference was Buzz Aldrin, the second man to step on the Moon and Neil Armstrong's Moon lander pilot. He was standing alone and I introduced myself. I was drawn to Aldrin for several reasons, primarily because he had been to the Moon but also because both of us shared a life experience: we both had probed the human frontier and then endured private turmoil as a result. In his book, *Return to Earth*, he chronicled his breakdown after his epic experience on the Moon.[4] After all of the potential his grand vision from the Moon had revealed, he had been unable to accept the fact that his mission to the Moon would have no immediate impact on humanity, and this disappointment had nearly torn him apart. However, without even realizing it, Aldrin had been instrumental in a monumental but little-known discovery on the Moon—a discovery that may yet save both the Earth and her humanity. Hope is sometimes discovered in strange places; in this case it was discovered on the Moon in an astronaut's footsteps.

The top few yards of the Moon's surface are a pulverized rubble that has been continuously overturned by meteoritic bombardment. The Moon dust which is now six feet underground was lying on the surface an eon ago. This process of savage overturning by repeated hypervelocity impact and explosion is called "gardening" by lunar scientists. While on the Moon's surface, the lunar soil is exposed to the solar wind—a blast wave of atoms and ions from the hottest outer portion of the solar atmosphere. The particles in the solar wind are traveling so fast they bury themselves like bullets in the dust particles at the lunar surface and are trapped there.

The Apollo astronauts took deep core samples while they explored the lunar surface. The dust from the core samples was supercharged with the gases that make up the solar wind. When they brought dust from the Moon back to Earth, Armstrong and Aldrin also brought back pieces of the Sun in the form of atomic particles, some of which were quite rare by Earth standards. In fact, one of these rare particles discovered in the lunar dust may allow humanity to safely harness the power of the Sun and the stars—and do it safely, without loading the atmosphere with any more carbon dioxide.

When the Moon's dust was heated, it released a gas that was rich in helium-3, an isotope of helium that is very rare on Earth. Helium-3 is made by the process of fusion in the Sun's core. From there it is lofted to the surface and becomes part of the solar wind. Helium-3 is non-radioactive and can be considered a portion of the Sun's fusion energy trapped in an atom. When helium-3 (He_3) is burned with deuterium (which is a stable isotope of hydrogen common on Earth), it forms ordinary helium and hydrogen and releases energy in the form of motion of the resulting proton. This fusion reaction is known as D–He_3 fusion and is the favorite of all fusion scientists like myself, because it starts clean and ends clean. Its fuel is non-radioactive, as are its end products.

While D-He_3 has been a favorite of fusion researchers, it was not considered practical, since there was little He_3 on Earth. For this reason, when a fusion researcher at the University of Wisconsin heard over lunch one day that his lunar science colleague had found an abundance of He_3 in the gas released by Moon dust, this was cause for rejoicing. When I heard this, I got down on my knees in my private office and thanked God for his deliverance of humanity. The discovery of He_3 resources on the Moon makes fusion far more viable as a power source on Earth.

Fusion is the form of energy that heats and lights the universe. It is the manifestation of a yearning of all hydrogen in the universe to condense into heavier and more complex elements, ending with iron. Fusion is the source of sunlight and stardust, slow, reliable, beautiful—nature at its most nurturing. And although travel to the Moon and Mars may not solve Earth's problems directly, it is surprising to find that, indirectly, it may. The most wonderful part of any journey is the sight you never expected to see.

In the early morning of April 26, 1986, the night shift came to work at Karlskrona Nuclear Power Center in Sweden, as they had for years. Later, as they left the plant and the day shift workers began to enter, the radiation detectors suddenly blared, alarms rang and the gates closed. Radiation had been detected by the sensors at the gates, thus triggering everyone's worst fear—that an unseen radiation leak at the plant had contaminated some of the night shift workers.

As the plant safety personnel scrambled to find the leak they were stunned to find that it was not the night shift that had triggered the alarm, but the day shift coming. The radioactivity was coming from some massive source outside the plant and it was exposing the public to levels of up to a hundred times the normal background radiation.[5] This meant that the problem was not some small concentrated leak within the plant that could be contained and stopped. Rather, it was a catastrophe somewhere out there in the larger world.

As the urgent search for the radiation source began to spread in Sweden, similar alarms began to sound all across Central Europe. A massive radioactive cloud was moving in on a weather front from the East. Even as the governments of northern Europe began to grasp the terrible scope of the cloud and wonder desperately what was its source, workers in the classified regions of the Pentagon were staring in stunned disbelief at the cause.

During the night of April 26, the super-secret listening posts and satellites of the Western Allies had been monitoring a curious increase in coded communications traffic in the western Soviet Union, an increase that had soon mushroomed and led to the furious mobilization of civil defense and emergency radiation crews—all now racing from their bases across the Soviet empire and converging on a place called Chernobyl.

What was at Chernobyl? This became the question of the moment, and soon came the stupefying answer: a large nuclear power station. Even as that answer came, the steerable photographic reconnaissance satellites were now beaming down their terrifying images. The meaning of the mobilization was now evident: there was a disaster at Chernobyl. And from the images and reports of radiation clouds spreading from the site, it looked to be a full-scale catastrophe.

As dawn came and the satellites focused their cameras and their infrared and radiation sensors on the Chernobyl nuclear station, a horrifying and heart-rending spectacle unfolded. The Russians were fighting several large fires at the plant, but in the center of the disaster was a massive and impossibly intense fire that was so hot, water was useless against it. Water from the hoses of the firemen was vaporizing in midair before it reached the flames. The fire was so full of radiation and toxic chemicals that its smoke would kill whoever came near, yet the firemen pushed forward to try and bring it under control. This was the inferno of a fission reactor core made of graphite (a form of carbon denser than coal) that had undergone an explosive meltdown. The fuel and radioactive waste and the graphite itself were burning with a combination of searing radiation and blast-furnace heat. A fleet of heavy military helicopters arrived, and, in a display of desperate heroism by their pilots, flew directly into the deadly smoke plume and dumped chemicals reported to be a toxic mixture of lead and sand on the fire's core.

All day the battle raged, and through the night and into the next day before the fires were brought under control. The true death toll has been kept secret, or is perhaps simply unknown, but a common report around Washington in the days following said that hundreds of firemen and pilots had died from either chemical poisoning or lethal radiation doses. Greenpeace estimates that at least 400,000 have been forced to leave their homes in Belarus, Russia, and Ukraine, and 270,000 people still live in areas with such high levels of contamination that there are strict controls on the use of locally grown food.[6] The World Health Organization and various governmental bodies have been monitoring the devastating health consequences, particularly for the thousands of workers involved in putting out the fires and other recovery work. Over 160,000 square kilometers (almost 62,000 square miles) are contaminated at a level that will take up to 130 years to recover.[7]

The fire finally out, tunnels were dug under the still-white hot core and filled with liquid nitrogen piping. This was the only way the Soviets could figure to stop the molten core from melting through the earth into the water table—the "China Syndrome"—which would cause a massive thermal explosion in which most of the plant's radioactive material would be blasted into the atmosphere. The ruins of the entire reactor building were encased in a concrete "sarcophagus."

But, even with China Syndrome averted, it was way too late already. A huge radioactive cloud spread over first Europe and then the entire northern hemisphere. Both the size and height of the cloud testified to the violence of the initial explosion. All over Europe immediate precautions had to be taken to ensure the food supply was not affected, and in many areas whole populations took doses of iodine to saturate their bodies and block the absorption of radioactive iodine from the cloud. The number of people suffering long-term effects or premature death from this exposure can only be estimated, but will probably be in the millions. One casualty is certain, however—the faith of the industrial world in nuclear fission.

Much of Europe had been pro-nuclear even after American faith in nuclear power had faded, along with trust in its government. To Europeans, even Three Mile Island (TMI) had been a regrettable but avoidable incident. Many rightly pointed out that TMI had actually showed the safety of nuclear power, since the incompetence of the reactor operators there had been so great that they could not have done more damage if they had deliberately set out to melt down the reactor core. TMI had not become a Chernobyl because it had a heavy containment vessel that had been built to withstand the pressures.

Amazingly, at Chernobyl, where so many died, the reactor operators responsible for the explosion survived, like the drunk driver who is the sole survivor of a wreck that kills a whole family. They had stumbled out of the wreckage and been arrested, then questioned on how the catastrophe had been triggered. When their story was heard they were immediately put on trial for negligence. What had spilled from their lips was a story of almost unbelievable stupidity and recklessness, which, when combined with the abominable design of the reactor, had led to the catastrophe.

There is a saying among sailors: "Nothing is fool-proof. Fools are much too clever." Folly will transform even the safest of situations into a disaster. So for a situation which is a disaster waiting to happen, it was particularly likely. The Chernobyl reactor number 4 was a graphite moderated reactor, one of the most primitive types of nuclear reactors, and as reactors go, criminally unsafe in design. Graphite, mixed with the fuel, was used as a major part of the reactor core because it was the first substance demonstrated to encourage a nuclear chain reaction in natural uranium. Ordinarily, the neutrons released in fission reaction come out too fast to trigger additional reactions efficiently, So in order to slow them down, the fuel is encased in a moderator such as graphite, which is one of the less precisely controllable materials. In a power reactor in the West, the moderator is also the coolant, normally water or high pressure carbon dioxide gas. This way, if the coolant is lost, or even if the reactor overheats, the chain reaction stops. The reactor shuts down, and produces heat only from decay of nuclear waste. At TMI, it was demonstrated that this could still lead to a meltdown. However, at Chernobyl, the cooling water actually over-moderated the chain reaction, so that when it was gone, the chain reaction in the reactor core was accelerated, and it became a runaway chain reaction. The reaction would build up again and again, still nicely moderated by the graphite, and would superheat until the graphite itself vaporized. Unfortunately, graphite is one of the most temperature resistant materials known, so the fuel vaporized first. The Chernobyl reactor then embarked on a runaway chain reaction until it exploded like a bomb, releasing in total radioactivity many times the combined releases from the atomic bombs dropped on Hiroshima and Nagasaki, although the specific radionuclides released were different. Given the conditions in the core at Chernobyl 4 that night, the whole process took only a second.

One consequence of Chernobyl is that the risks of nuclear power are no longer considered acceptable in much of Europe. Given that the radioactive dust from Chernobyl can still be detected in the glaciers of Switzerland, this is an opinion that is well founded. What is needed is a new source of power that is much safer and cleaner than fission, but which can still provide large amounts of power for the growth of a just, environmentally-benign global civilization. Fusion power is such a source. Yet, between the mistakes of Chernobyl and Three Mile Island and the haunting threat of nuclear weapons, can the hope of fusion, a far safer nuclear technology that is unsuitable for weaponry, be redeemed?

On St. Patrick's Day in 1986, on the top floor of a 30-storey office building, we began full-scale thermonuclear war. This was a full-fledged war game between the blue and red teams of SDI, with a gold team of judges observing. The gold team filed in and took their places—dour old men, retired generals and admirals, grizzled scientists and engineers, veterans of Cold War combat and open-air nuclear tests. They were a grim, tough, wizened-looking bunch.

The red team, including myself, were simulating the Soviet Union. We began the war by launching a surprise attack on the United States, which had in place an early, thin version of a Star Wars system. Our furious attack concentrated on the SDI system with a variety of space- and ground-based weapons which were known to be in the Soviet arsenal. This triggered a massive war in space as the SDI system defended itself. What remained of the system after five minutes we atttacked with nuclear-tipped ground-based interceptors. Outcomes were figured based on probabilities—a sort of Monte Carlo meets the end of the world.

Only after dozens of nuclear explosions had occurred in low orbit over the Soviet Union did we launch our main attack using the massive SS-18 ICBMs to try and decapitate and cripple the U.S. nuclear forces. Despite our attack, some of the SDI system survived and some of our SS-18s were lost, but we had compensated for this by hitting high-priority targets multiple times. As we all watched and drank our coffee, the strings of warheads impacted on viewgraph after viewgraph of the United States, showing the evolution of the war in its 40 minutes of surreal time. The war lasted all morning, and then broke for lunch and continued into the afternoon. When it was done, the world was utterly destroyed. It could be said that surprise and certainty were the only casualties the SDI system had inflicted on our attack.

Remarkably, there was a long discussion by the gold team over the results. While everything that meant anything to humans was totally lost, the discussion became a heated argument concerning our assumptions about the Soviets' relative mix of ground bursts and air bursts to destroy American missile silos. One grizzled expert after another rose to contribute to this debate. There were discussions of ground- and air-shock reflections and overpressures and Mach stems. I sat in a daze at the end of the day and then joined several friends at an Irish bar for a drink. It was my first experience of the effects of a nuclear war and I felt as numb as its last survivor. I had to be carried out of the bar when we left, perhaps an excessive reaction to what I had seen that day, but somehow appropriate. I arrived throughly hungover the next morning for the second day of the meeting.

On the second day, the heated debate over ground versus air bursts resumed, and continued ferociously for the rest of the day. If one ignored the words, it was a prolonged cry of anger and dismay over what had been witnessed, yet the ongoing fury clearly denied the obvious. There was nothing left to fight about.

CHAPTER 7

Life Rules

Mars sent Earth a love note 16 million years ago. Inscribed on a stone, toward the starry ether thrown, it took millions of years to be captured by gravity to take a flaming plunge through the currents of our planet's atmosphere. Fallen and alone, it lay on ancient ice unread for 13,000 years.[1] But, when its code was broken and its legend read; the message that was etched in stone told humans we are not alone. "I too have life."

It had been several years since it had been established that the meteorite EETA79001 had originated on Mars. When that was determined, the whole category of meteorites, the SNCs (Shergotty, Nakhla, and Chassigny), were established as Martian, too. One of the characteristics this group of very young meteorites was that they shared a similar pattern of oxygen isotopes. In fact, if you were to chart these, they would all fall on the same line, known as a "fractionation line." Any meteors whose characteristics put them on this line would be recognized as Martian as well. Rocks from Earth, based on their oxygen isotopes, also fall on their own fractionation line, yet it is a very different line on the graph from the Mars line. We can clearly differentiate Mars rocks from Earth rocks in two ways: chemically, based on their oxygen isotopes, and visually, based on where their isotope readings fall on a graph.

The very fact that the SNCs were so young posed an interesting problem, a problem the Mars researchers called the "age paradox." Mars was a planet whose surface was old. In the Lunar Mars model it was on average 4 billion years old or more, yet the SNCs were all dated at only 1 billion years old or less. Assuming, reasonably, that the SNCs were from various sites on Mars and ejected into space by many different impact events, the meteorites should have represented a wide variety of ages. So why was Mars only sending us young rocks? Something was missing. Where were the old Mars rocks?

These Mars meteorites were paradoxical in another way, too. They were telling us something else, something that we had been trying to distance ourselves from, something the scientific community was having a great deal of difficulty in accepting: they told us there was organic matter on Mars.

The Viking mission had supposedly already told us that there was no organic matter on Mars,[2] but in the water-deposited limestone that was sandwiched between the melt glass of EETA79001, incriminating evidence was discovered: organic molecules. Here, on a rock from Mars that had traveled all the way to Earth, were easily detectable amounts of organic material right next to carbonates. This strongly suggested that Mars, like the Earth, may have been the home to lifeforms that absorbed carbon dioxide and deposited it in sedimentary rocks.[3] This was a great discovery! Call the press! However, this was also a big problem in a number of ways, so no one picked up the phone.

First, there was a major scientific problem with this result. The evidence of life was discovered in the youngest rock from Mars—EETA79001, now the Rosetta stone of the Mars meteorites. EETA79001 was only 160–175 million years old, dating it back to the age of the dinosaurs on Earth. This discovery meant that—if the evidence of life was confirmed—Mars had not only once had life, but its life had been growing on the surface of Mars until, geologically speaking, extremely recently. So recently, in fact, that it could be argued that it had lived until (almost) the present time. If not, well . . . we begin to see what the problem is.

Remarkably, organic matter in large amounts had been found in meteorites long before EETA79001 was discovered. Some meteorites were thick with it. These meteorites, then of unknown origin, are known as carbonaceous chondrites. They are of various types, distinguished, as were the Mars meteorites, by oxygen isotopes. The ones that had organic matter in most abundance are called CI or C-1, and are very rare. The most common type are called CM or C-2, and they are also loaded with organic matter.

It was not unexpected that organic matter could be found in meteorites that were 4.5 billion years old. The standard model for the formation of life on Earth was that organic matter was either synthesized "locally" or, as an alternative, that life rained down upon our planet from above. Everywhere astronomers looked they found organic matter in space, vast clouds pregnant with organic material wandering our universe. Might not such clouds have congealed into meteorites when the solar system was forming? But such speculation again raised the awful question: was humanity the ruler of the living universe or just a speck of dust in a cosmos full of living things?

What was even more disturbing was that other things were found in the meteorites that weren't merely suggestive of life but had direct evidence that something had lived in them. Microfossils had been reported in the CIs—lots of them. There was no mistaking them. They would have been obvious to Robert Koch himself had he examined them under his microscope in the early 1800s. It was as if colonies of bacteria had thrived in them, then died, leaving their bodies to science.

The microfossils aboard the CIs had been found in the 1960s by a fellow named Bartholomew Nagy, a true gentleman and scholar.[4] He had modestly announced his results and was thereafter assaulted by a mob of scientists trying to prove that he had found only terrestrial contamination. The mob formed quickly and, with amazing vehemence, accused him of finding Earth bacteria and thinking it was from outer space. (This is a theme that continues to be repeated today, even as the evidence of biology aboard meteorites continues to mount.) He fought back doggedly and ably defended his work, although he was by nature a reserved and inquiring individual. However, by then, after the initial announcement and flurry of attacks, the press and the scientific establishment had quit paying attention.

Science quickly wearies of controversy. Controversy breeds animosity and animosity clouds the mind. Clouded minds produce poor science and, after a few years of fierce brawling, everyone moves on. Nagy was lucky in a way. The Inquisition burned Giordano Bruno at the stake for claiming there were other worlds with other peoples on them. Nagy published one final book on his work in 1975, entitled *Carbonaceous Meteorites*, then moved on to the problem of finding microfossils in places where it was safe to find them: in old Earth rocks. He was able to live out his life in relative peace, and his work on the CIs lives on, having profoundly influenced the emerging field of exobiology.

I was transfixed by the Martian "age paradox." I knew that the whole problem of Mars' surface ages versus its meteorite ages boiled down to one number. That number was the controlling parameter of Mars' geologic history: the rate at which craters were created. If Mars had a high rate of cratering, then its surface was young and so were its meteorites, which would be evidence to support the conclusion that the SNCs were indeed from Mars. However, even if the cratering rate was high, the southern highlands of Mars still looked as old as the ancient plains on the 4.5-billion-year-old Moon. So while indeed there could be young rocks from Mars, younger than those found on the Moon, logically there still had to be old rocks from Mars as well. Where were they?

I went to a conference and presented a paper where I announced that the Mars cratering rate was at least four times the lunar rate.[5] I was concerned that this hypothesis would be disturbing to the other scientists, but the new notion was well received. Everyone grasped that if the Mars cratering rate was allowed to be four-times-lunar, the age paradox disappeared.[6] But, I said, an old Mars rock had to be found for my thesis to be correct. Little did I know that such a thing, 4.5 billion years old, had already been found.

I saw the researchers who'd discovered the organic matter in ETA79001 and asked them why they hadn't found any more. They replied they were "scared to find more." The reaction to their original finding had been so hostile that they had quit looking. To look for signs of exobiology was to look for trouble, and they weren't looking for any more trouble. I was furious. "It takes courage to discover things!" I insisted, finally quoting Lady Macbeth, "you must screw your courage to the sticking point," to demonstrate what must be done.

On the plane home I was seething. As was typical of one weakness in the scientific method, many scientists were finding evidences that would lead to great discoveries, but through caution and timidity were simply not reporting them. They didn't want to be subjected to the arguments that might be mounted against their findings. Earlier this century, Wolfgang Pauli had been notorious for talking young physicists out of publishing their discoveries, only to have someone else, someone outside his circle of influence, publish them. Uhlenbeck published his discovery of electron spin after Pauli had browbeaten another young scientist, Kronig, out of publishing the same evidence.[7] Pauli did the same to Stueckelberg, who'd discovered the meson. This discovery was instead published by Yukawa.[8] Some people lived in fear of being wrong in Pauli's eyes, while others considered him simply an acerbic bully who just happened to be a genius. Nevertheless, evidence of life on Mars was not being pursued because it was professionally uncomfortable, and I wasn't happy about this situation. A finding of life from Mars was too important and I was determined to force these reluctant "heroes of science" out into the open. Besides that, I had my own hypothesis to defend and I was still looking for an old Mars rock to defend it.

᾽ I had read of Nagy's work with the CI family of meteorites and decided I would check their properties to see if perhaps they were from Mars. It seemed a remote possibility, but sometimes meteorites have been misclassified for years. So I returned home and quickly devoured the literature on what makes a meteorite Martian. It was the oxygen, I concluded. If the oxygen was heavier than Earth's, but not much heavier, then this identified the rock as Martian. There were a host of other tests, but this was the most important—the deciding one.

I went to the local university and scanned the books in the library. One new book on meteorites showed a graph of oxygen isotope ratios and where the different types of meteorites lay. Everything on Earth lay on a straight line, a "fractionation line" on these charts, and everything on Mars lay on a slightly higher but parallel line. Many other types of meteorites were also placed on the map, but where were the CIs? I knew the oxygen isotope pattern was the crucial question. I was almost shaking with tension as I scanned the graph. If the CIs were Martian, the question of life on early Mars would be settled in one day. If they were from Mars, their oxygen would have to be on the

Martian line, but the CIs were nowhere to be seen on the chart. Then I turned the page, where there was a much larger graph. There, finally, I found the CIs. They were way up along the Earth fractionation line all by themselves, and as my heart soared, I saw they were slightly above it! They were on the Mars line! I rejoiced; Eureka! I'd found it! The CIs were from Mars. The CIs were 4.5 billion years old, and they were thick with microfossils! They were also composed of hydrated clay, like that from an old, dried-out lake bottom.

There really had been life on Mars, and I had discovered the confirmation. Unlike Archimedes, I did not run naked through the streets of Washington. Although I was probably just as happy as Archimedes had been, I have other, less revealing, ways of showing it.

As with so many scientific discoveries, when you begin to investigate one thing, you catch a whiff of other interesting phenomena. In my exploration of Mars' cratering rates I had noticed something unusual. I had noticed that the rate of water erosion on Mars dropped by a factor of 30 between the Late Hesperian and the Early Amazonian ages on Mars, which probably ran between 1 and 0.3 billion years ago. Mars' epochs are named after prominent regions of that age: Hesperia is a vast plain in the south of Mars, covered with dried water channels; Amazonia is a region of frozen desert plains stretching from Elysium to Olympus Mons, smooth and almost devoid of old water channels. What had happened during the transition period between the Hesperian and Amazonian epochs, I wondered. Why was Hesperia so water carved and Amazonia so barren?

We soon found out. A large crater had appeared on Mars in the Early Amazonian age. In fact, this crater was not only the last large crater formed on Mars, but the biggest crater to have formed on Mars since the epoch of heavy bombardment. The crater had been named Lyot (pronounced "Leo," from the French) and is 120 miles across and lies 300 miles northeast of Cydonia.[9] Imagine, Mars had peacefully passed most of its history with no large impacts, its early violent period long forgotten, then BOOM. It was as violent and stupendous a strike as the impact at Chicxulub in the Yucatan had been when the reign of the dinosaurs ended.

What would the life cycle of Martian microbes be like? Would they lie dormant, waiting for the rare day when temperatures rose above some magical points then burst into a flurry of microbial activity? Or would they slowly process the meager solar radiation that arrives in their vicinity in a slow-motion photosynthesis?

Gil Levin, the scientist whose on-board test during the Viking missions had produced positive readings for life on Mars, is the world's leading authority on Martian microbes in the new field of exobiology (a combination of astronomy and biology). He has achieved this de facto honor after spending the last 20 years doggedly disproving the firestorm of scientific objections that were raised to his claim that "the LR [Labeled Release experiment] data are consistent with a biological answer."[10] In fact, Levin has disproved every single serious objection raised about his findings. In doing so, he has begun to unravel the biological processes of Mars—methodically, precisely, and scientifically. His extensive work supporting the assertion that Mars is biologically active is now the subject of a book, *Mars, The Living Planet.*[11]

Everywhere he goes, Gil Levin tells audiences that his test demonstrated that there are at least minimal biological processes occurring on the surface of Mars right now. To date, though many have tried, no one has been able to prove him wrong. If anything, the arrival and confirmation of microbes aboard the Martian meteorites would seem to add credibility to his case.

Here on Earth, microbial life is ordinarily symbiotic: Earth microbes frequently "live together" with other lifeforms in a relationship of interaction and codependence. The interaction between species can be very simple or very complex, and beneficial to one or both, or detrimental to one for the benefit of the other.

There are four fundamental types of relationships between symbionts. The first two, commensalism and amensalism, are relationships in which the different species live in close proximity to one another. They "eat at the same table," so to speak. In commensalism, one of the members is unaffected by the relationship and the other benefits from it. In amensalism, one of the members suffers as a result of the relationship and the other is unaffected.

In the other two forms of symbiosis—mutualism and parasitism—one member of the relationship is totally dependent upon the other, the host, for nutrients, habitat, and other life functions. In mutualism, both of the species involved benefit from the relationship. In parasitism, the dependent member benefits from the relationship, while the host is harmed by it.[12]

Symbiosis is, at the very least, a useful metaphor for humanity's relationship to Earth. While symbiosis is usually a strictly biological consideration, we can easily think of Earth and its biosphere as a single living organism. Although we are not microbes, humans *are* symbionts, and we are utterly dependent upon Earth and its biosphere for everything that we need to survive. Our host must provide us with habitat, food, water, shelter, warmth, and the respiratory gases we require. So we are in a symbiotic relationship in which we are 100 percent dependent. Therefore, we can be in one of two relationships with Earth. We are either mutualists, in which case both humans and the Earth's biosphere derive benefits from our relationship, or we are parasites and derive benefits at the expense of other life on Earth.

If you were to look back upon the Earth from another planet, say from Mars, and were assessing humanity's relationship with its host, what would your assessment be? Does the Earth derive net benefit from its relationship with humanity, as well as the other way around? If so, what benefit is the Earth deriving? Or, are we living here to the long-term detriment of Earth's biosphere? What would you conclude if you examined humanity's effects on Earth's air, water, other animals, and plants?

In a parasitic relationship, the host will ultimately be deprived—literally sucked dry—of what it needs to sustain its own life and the parasites will kill it. The parasites must then discover another host and colonize it before the new host dies, or the parasites themselves will die, and if they are a single colony, become extinct.

Peter Price, author of *Evolutionary Biology of Parasites*, points out that if we choose to continue a parasitic relationship with our host, we have chosen an evolutionary dead end. "Parasites as a whole are worthy examples of the inexorable march of evolution into blind alleys."[13] So the very choice to be parasitic may be retarding our evolution. Ultimately, if continued, this course of action will end in our own extinction. "The commonest view on the evolution of parasites is that they evolve slowly and represent dead ends in

141

any phylogeny."[14] Unfortunately, we're evolving rapidly—into excellent parasites—so unless we can quickly colonize Mars or change strategies, our demise will probably be equally rapid.

This observation raises some interesting questions. If we are parasitic now, have we always been parasites? Is parasitism intrinsic to our species? Are we "trying it out" and only now discovering that it leads to failure? Do we have the option of becoming mutualists? Do we remotely know how to live in a way that will "benefit the Earth"? Even the idea seems strange. Can we switch from one symbiotic strategy to another? Can we change before we kill off our habitat on the Earth? Can we change before we kill ourselves?

Humans do not have another obvious host. Is it possible that our intense drive and interest in exploring space is nothing more than following a genetically engineered instinct to locate and colonize a new host before the old one dies? We must look at our unconscious strategies and create a long-term plan based on conscious choice, because unless we change from parasitism, we are doomed.

Even Mars could not provide us with what we need to survive and reproduce—certainly not for tens of thousands of years—and through our exponentially increasing damage to the Earth, it is much less than certain that we will make it to the end of the 21st century.

If we do survive, it will be because we've fundamentally altered our relationship with Earth. We'll have to profoundly modify our survival strategies. Hardest will be those that we are inordinately fond of, those that we are convinced are "modern solutions," and those that are beneath our conscious awareness. Clearly, it will require hugely creative thinking, ingenuity and diligence to invent a mutualist future for humankind. Such a future will have to be almost entirely invented, because right now even our most "effective" solutions are barely slowing the approaching juggernaut of planetary death.

Space research and exploration may give us that chance. From a distance, as the Bette Midler song says, it is so much easier to see both nature's beauty and its absolute indifference. From a distance, Earth is the sole living sibling in a family of dead planets, a shining dot in a great black sea. It has a single biosphere, one linked ocean and a seamless sky. This viewpoint may save humanity. Both the ozone hole and the extent of rainforest destruction in the Amazon were first recognized from space. From "out there" we see that Earth is a planetary oasis among many planets which cannot sustain life as we live it. We reside on a living, breathing orb in a vast wasteland; a planetary ark on a cold and indifferent sea. It is said that the environmental movement was triggered by the magnificent image of Earth rising above the Moon taken during the Apollo program. Only from space can we see how limited the Earth is, how small and fragile. From a viewpoint in space we can more clearly understand how much trouble we are in. Mars is essentially a planetary "Exhibit A," warning us that the cosmos has no safety net, that the worst things imaginable, the deaths of entire worlds, can and do happen. Now that we have this knowledge we are responsible for acting appropriately. Already the damage to Earth has advanced beyond local environmental concerns—a polluted river or metropolitan airspace—to the point where we can now recognize that all effects and all possible remedies are indivisibly planetary in scope.

If the Earth's biosphere was an organism and its vital signs were being monitored, one would conclude that the Earth is dying. Such vital signs would be: levels of biodiversity, quality of ocean habitats, patterns of global climate, the balance of atmospheric gases. All of these, and more, are in decline. Not only is the carbon dioxide level going up at an average rate of more than 1.5 parts per million a year, but oxygen is declining at an even faster rate of almost 3.8 parts per million a year.[15] However, not all signs of biological distress are seen from space. Some we can see face to face on Earth.

He stood stock-still and covered his face. He cried aloud on heaven to break the nightmare or to let him understand what was happening. A trail of mutilated frogs lay along the edge of the island.

C. S. Lewis, *Perelandra*

Amphibians dwell at the interface of air and water, having skins in some cases so porous that they perform a major part of the animal's respiration. They begin their life cycle as eggs floating near the water's surface, exposing them to whatever is in the air or water. Consequently, they are extremely vulnerable to human-produced environmental toxins, especially pesticides. Their numbers appear to be dropping precipitously and, in a dreadful warning straight from some bad science fiction movie, large numbers of mutated frogs have now appeared in widespread areas in the Midwestern United States.[16] Given the rural setting, the culprit is most likely some form of pesticide or increased UV-B radiation, the result of damage to our ozone layer, or some terrible synergism of both. The malformed frogs, crippled by multiple extra legs, with extra eyes gazing at us sightlessly, are unprecedented and cry out that something is going terribly wrong. Predictably, there are scientists who claim that this is merely some fungal infection that no one has ever reported before, hence if we ignore this problem, it will go away. They are the pathologically skeptical, but in this case skepticism can be deadly. The belief that skepticism carries no risk is a fallacy of the "brownlash" movement that seeks to minimize concern over global change.[17] The truth is that all attitudes, those requiring action as well as inaction, carry risk, and the risks themselves cannot be assessed with certainty because the necessity for decision always outruns information.

It is not a time for seeking certainties, it is a time for estimating risks—the probability of human complicity in global change, multiplied by the damage that will accrue if we do not desist. Since the potential damage is terrible, even a low probability is enough to argue that certain activities should cease. There are scientists who insist that all the effects, from deformed frogs to ozone holes, to, finally, greenhouse gas-induced global warming trends, are either illusory or misunderstood natural effects. How certain can they be? Given the potentially life-threatening consequences, shouldn't we err on the side of safety? Having identified the probable human causes, isn't it better to dramatically curtail these activities? If we are cautious, even overly cautious, little damage will ensue other than a slowing down in economic growth. But if we fail to act, to act conservatively—to conserve life on Earth—then the real price in catastrophic economic losses could bring the U.S. economy and the world's economy to their knees. British Environment Minister Michael Meacher has suggested that, "People are starting to wake up to the cost of devastating climate change." He warned that the economic costs of rising seas, hurricanes, flooding and heat waves will "dwarf the costs of trying to prevent them."[18]

It is reasonable to ask how problems such as global warming can be remedied with the minimum of economic detriment and dislocation, especially for the poor. Economic dislocation kills people as surely as does pollution or climatic change. But it is also reasonable to consider that, while the economic costs of environmental problems accrue to everyone, as with most environmental problems, the burdens fall disproportionately on the poor. Yet, economic consequences can cut both ways, since there may also be economic benefits for those companies and organizations that innovate and develop new energy solutions, as Amory Lovins at the Aspen Institute points out so powerfully. Nevertheless, fears about the economy do not represent an adequate justification to delay solving the problems of carbon dioxide. If we are truly committed to a vibrant world economy, the best strategy would be to make an all-out effort to ensure that safe, low-cost, low-carbon energy is available to everyone.

Earlier in this book we talked a bit about the interaction of plants and animals in a symbiotic exchange of oxygen and carbon dioxide. This, of course, is just one aspect of an exquisite interaction between all of the plants, animals, and elements of our planet. One of the grandest of these is the relationship between the atmosphere and the ocean. Embracing like eternal lovers, the ocean and the sky mix and fold into each other, blending into a system of currents and chemical exchanges from which they can never separate. What is done to one will both directly and indirectly affect the other.

Most of us are aware of the fact that the forests around the globe, particularly the rainforests, are considered to be major "sinks" for carbon dioxide. (A sink is a place where one substance is absorbed, stored, processed, or transformed by another substance.) While the forests of the planet are essential to the processing of carbon dioxide, the ocean is another and far greater carbon dioxide sink. ". . . the ocean is the largest 'sink' for man-made carbon dioxide today and in the long run, is the place where about 90 percent of that carbon dioxide will end out," explains Jorge L. Sarmiento of Princeton University's Fluid Dynamics Laboratory.[19]

One of the significant areas of scientific research today is the study of the global carbon cycle, including the capacity and location of carbon dioxide sinks. Many scientists are engaged in the task of discovering where the various carbon sinks are, measuring how much carbon those sinks can hold, and considering what processes might change their capacities. In support of this effort, carbon samples are being taken from such diverse places as the Arctic ice fields, the upper atmosphere, and tropical rainforests.

Another major aspect of this inquiry is the attempt to statistically and digitally model carbon dioxide's interaction with plants, land, water, and atmosphere. The task of creating computer models which accurately reflect all of the natural processes, including such features as ocean currents, cloud cover and seasonal changes, has proven tremendously challenging. However, there is the hope that such a "model" will be able to help us test various scenarios in order to more readily understand how carbon dioxide affects our climate and the ways in which our planet may react to ever-increasing carbon levels. This is a worthy goal which has had the unfortunate side effect of generating scientific debate on the merits of one model over another; even worse, there has been a tendency to discredit the findings of global warming as a general criticism of the shortcomings of the models. This is a serious mistake. We certainly don't have to wait until the computer models are perfected in order to see the effects of global warming. We are surrounded by them. However, both global warming and carbon dioxide are invisible and we tend to relate to them as concepts rather than physical realities. So we are in need of a scientific concept that accurately represents our circumstances.

Here is the critical concept, so simple that it might easily be overlooked, yet it reflects the single key point which must not be lost in the argument over very complicated planetary models: Our planet is in a state that we are calling "All Sinks Plus," a state in which we are putting carbon into the atmosphere at a rate that is greater than the sum total that "all sinks" can process or absorb.

We can assume that we are exceeding the capacity of "all sinks" because there has been an increase in atmospheric carbon dioxide every single year, year after year for at least the last 150 years. It is therefore evident that "all sinks" are presently absorbing at full capacity. Even if the capacity of "all sinks" has expanded over the years, it falls short of the levels of carbon being put into the atmosphere.

146

Right now, the rate of increase is about 1.5 parts per million per year. Twenty years ago, it was about 0.8 parts per million. The question is not whether carbon dioxide is increasing, but, in any given year, by how much the carbon dioxide released by burning exceeds the capacity of all oceans and land-based carbon sinks to absorb it. Effectively, as long as there is an increase in atmospheric carbon dioxide, we know that "all sinks" are fully utilized. There is no further resource to tap, or if there is, we do not know what that capacity is or how to utilize it. So the atmosphere is where the "plus" in All Sinks Plus must stay.

It is as if the water flowing from the tap into your kitchen sink is flowing in at a rate greater than the sink's drain can remove it. The water has already filled the sink and is flowing over onto the floor. You don't have the option of reaching in and opening the drain because it is already open. The water can't flow through the drain any faster.

If our kitchen were the environment, here is how we would be relating to it now. When we were wading in water up to our knees we would look into the sink and see some sponges floating in it. We know that sponges absorb water so we'd set up a team to study the sponges.

A "Save the Sponges" movement forms which hopes to help the problem by making certain that we will have an adequate supply of sponges. Another scientific team reasons that since we don't actually know the capacity of the sink, drain or sponges, nor where the water is coming from before it comes out the tap, how can we be sure that there is really more water pouring into the sink than can drained away? Measurements are called for and a research project is initiated. Yet another group is monitoring the rate at which water is flowing out onto the floor.

Meanwhile, various other studies are attempting to digitally model the dynamics of the kitchen. These models are being ridiculed because they haven't taken into consideration the possible tidal action of the Moon. The makers of dish soap, concerned about future earnings, release a finding that a well-funded in-house scientific research group has discovered that the moisture on the floor may actually be caused by normal cycles of wet and dry. Water may actually help to keep the floor clean!

The mop manufacturers agree that this is a natural wet cycle and recommend that in order to keep kitchen mops affordable and to make certain that jobs are unaffected, a new mop head should be purchased every month or two. A new industry-approved "maxi-mop" line is introduced to help cut down on the wet floor problem.

The *Action Evening News* team, obligated to present "both" sides of the story, presents a film on the ten o'clock news of you, the homeowner, watching helplessly as the sponges and your kitchen furniture float around while a dish soap scientist claims that there is no problem really as long as you buy a new mop.

We hope you will forgive the overblown metaphor, but the intention here is to illustrate our predicament. Sometimes the drama we are enacting can only be seen by exaggerating to make the point. Nevertheless, concern over carbon sinks is very real. As biogeochemist Richard Houghton of the Woods Hold Research Center in Massachusetts suggests, "If you understood the mechanism, you'd be in a much better position to say whether the sink will continue."[20] Commentary in *Science* agrees: "Whatever is going on, researchers are eager to sort it out before one or more of these mysterious sinks stops working."[21] The clear implication is that we don't really know what we are doing—and that is not good.

E very time we hear the cork popping on a bottle of champagne, we know that as likely as not a frothy foam of the bubbly liquid will spew out the bottle—unless, that is, it is being decanted by a wine steward who is skilled enough to slowly release the compressed carbon dioxide. Children rejoice when they discover that if they shake a can of a carbonated beverage, when opened, it releases a wonderful, cold spray guaranteed to upset the average adult.

Carbon dioxide is easily assimilated into water, where it forms carbonic acid, giving carbonated beverages a slightly tart taste. Carbon dioxide is just as easily released back into the air, however. It is clear, from the example of the champagne bottle and the can of soda pop, that if the carbon dioxide solution is under pressure and there is any percussive action, it is very happy to escape from solution and rejoin the atmosphere with a vengeance, taking half the can of liquid with it in the process. So, while under pressure, carbon dioxide will remain in liquid solutions in high concentrations, but it isn't stable. Drop the can to the ground and it becomes a bomb. That glass of carbonated liquid, which under one set of circumstances will slowly release gentle bubbles into the air, will behave quite differently under another set.

A young woman of our acquaintance, a mischievous ten-year-old, loves to horrify her mother and her friends by shaking a cold can of soda and threatening to spray the contents all over everyone. While her mother is yelling for her to stop, the ten-year-old smiles and pops the can top, which emits only the normal whisper of air. "How did she do that?" we wonder in amazement, relieved that we aren't mopping up from a fountain of spray. "Like this," she explains, shaking another can, "watch." Just before opening it, she briskly taps the top of the can a couple of times with her index finger (she claims you can hear when it is quiet inside) and when she pops the top, we witness another miracle: nothing happens.

Clearly, carbon dioxide goes into and out of solution easily. One minute we are expecting it to suddenly leave solution in a race to enter the atmosphere and then it happily goes back into solution with a slight tap. The next minute, when all seems calm, that same slight tap might suddenly send a torrent emerging from solution. And, if those circumstances are unfortunate enough, the results can be disastrous.

In an isolated village in western Africa, a tiny miracle occurred in the midst of a huge tragedy. When rescue workers arrived on the morning of August 22, 1986, they found a newborn infant crying between the legs of her mother who had died during the night while giving birth unattended. The poor mother was not the only one to have succumbed during the night, however. By the time the rescue workers got to the remote location, nearly everyone in the village was dead. Ultimately 1,700 people and 19,000 animals died. They were all asphyxiated by a deadly toxin they had inhaled during the night. Virtually no one lived. The doctors who treated the infant still can't explain how she survived.[22]

Unfortunately, the entire tragedy might have been averted had more serious attention been paid to a warning that occurred a couple of years earlier. As warnings from the Earth go, it was a tiny nudge. A mere 37 lives were lost, which on the planetary scale of things was insignificant, although there is no denying how each of these people and everyone connected to them suffered. On August 15, 1984, Lake Monoun, in Cameroon in western Africa, discharged a large amount of carbon dioxide which previously had floated silently in a compressed layer just above the lake's floor. The released gas formed a cloud which drifted silently to shore and sucked the life from a small coastal village in a single breath. No one understood precisely how it happened, and a pair of scientists who urged that the cause be researched were not funded.[23] So, in the way that human tragedy is all too frequently forgotten, the incident was on its way to becoming an interesting footnote in geology textbooks: an anomalous incident involving carbon dioxide, nothing more.[24]

Lake Monoun is one of a series of lakes that nestle in a dense jungle region near the Atlantic coast of equatorial Africa. The lakes, many surrounded by beautiful jagged cliffs and cascading waterfalls, are at the center of a bucolic village life. Seven of these villages are situated around Lake Nyos, which in 1986 was known locally as "the good lake," and community life there was widely acknowledged as happy. It was a place where those who grew crops and those who raised livestock lived in peaceful coexistence, with their round, thatched and mud-brick homes held in the security of the natural valley surrounding the lake.

Lake Nyos, about 5,900 feet across, filled a maar (a round geological depression formed by an earlier explosion of carbon dioxide), at an elevation of 9,876 feet. It is estimated that this particular maar is approximately 400 years old, and it is assumed that the source of the carbon dioxide was past volcanic activity, although there have been no problems with volcanoes that anyone could remember. Analysis of the lake bottom confirmed that there had been no recent volcanic activity.[25] However, there were springs in the area that would percolate, as if boiling, from the carbon dioxide that was naturally found in the water.

On August 21, 1986, after nearly everyone in the sleepy lakeside communities had gone to bed, a large volume of carbon dioxide swiftly arose from the dark lake bottom where it had slowly accumulated, perhaps for years. It broke the surface of the water with enough force to splash water out of the lake and then formed a cloud containing approximately 7,000 of the 10,600 million cubic feet of the gas that scientists estimate had lain on the bottom of the lake, over 650 feet down.[26] The gaseous angel of death that gathered over the water then slowly spread, forming a toxic blanket that would suffocate and chemically poison every living creature in its path, and seeping into homes as easily as the night air. As the cloud slowly drifted, it consumed Subum, a lakeside village, in a single breath.

Because carbon dioxide is denser than air, it flows close to the ground, following the low places in the terrain. So the cloud didn't stop, but flowed over the crater rim and down a stream valley at a speed of some 65 feet per second, killing every person and every animal along a 16-mile path of destruction.[27] Ultimately, 1,700 people died, nearly another 1,000 were hospitalized, and 20,000 residents were displaced.

While it is hoped that processes have been put into effect to prevent this kind of tragedy happening again, there is good reason to believe that the warning of Lake Nyos should not be ignored, by any of us. It is easy to distance ourselves from its possible significance by reasoning that we don't live near lakes in volcanic maars or carbon dioxide laden springs. However, it is important to understand the mechanism involved so that we can, as it becomes appropriate, weigh the risks of a similar disaster on a much larger scale.

In the case of Lake Nyos, although it isn't absolutely certain, it would appear that the source of the carbon dioxide was groundwater that had entered the lake bottom, where it slowly accumulated over a long period. Volcanism was eliminated as a possible source since none of the other indications that are characteristic of a volcanic release of gases was present. Carbon-14 dating eliminated biogenic sources, so the carbon dioxide wasn't the result of plant life. By a process of elimination, it was assumed that the lake had been fed by carbonated springs, which are numerous in the area.

Once the carbon dioxide laden water entered the lake bottom, it stayed there in a stratified layer rather than mixing with the other water in the lake. This is both because of a temperature difference and a chemical difference between the carbonated bottom layer and the non-carbonated layers above. The cooler, denser, more saturated, and therefore heavier, carbonated water stayed on the bottom. The warmer water formed a cap on top, keeping the bottom layer's carbonated water under pressure. This situation may well have continued for a long time, allowing the pressure on the lower layer to gradually build.

At some point, this saturated layer became unstable. While it isn't clear exactly what triggered the release of carbon dioxide at Lake Nyos, it is clear that, given the circumstances, it could have been almost anything. For example, it might have been triggered by any disturbance that would cause deep water to rise to a point where supersaturation conditions were met.[28] The trigger might have been as innocent as a gentle breeze which caused the water on the surface to roil in a manner that broke down the lake's stratification layers and released the carbon dioxide. It is not uncommon for even large bodies of water—for example, the Great Lakes in the northern United States—to periodically "turn over," causing a reversal of top and bottom water. Once the release of carbon dioxide was triggered, anyone who has opened a foaming bottle of champagne or a shaken can of soda would recognize the results. It was as if a 7 billion cubic foot can of soda had just popped its top and spewed wildly into the air. Only this soda violated an important "life rule" for humans: we need to breathe oxygen. It thus caused death; death by soda pop.[29]

CHAPTER 8

The Vortex

In the spring of 1912 one of the largest moving objects ever created by human beings left Southampton and began gliding toward New York. It was the epitome of its industrial age—a potent representation of technology, property, luxury and progress. It weighed 66,000 tons. Its steel hull stretched the length of four city blocks. Each of its steam engines was the size of a townhouse. And it was headed for a disastrous encounter with the natural world. This vessel, of course, was the *Titanic*—a brute of a ship, seemingly impervious to the details of nature. In the minds of the captain, the crew, and many of the passengers, nothing could sink it.[1]

The world we know is like the *Titanic*. It is grand, chic, high powered, and it slips effortless through a frigid sea of icebergs. It does not have enough lifeboats, and those that it has will be poorly employed. The Americans, Europeans and Japanese are its aristocratic, first-class passengers. The Third World is in steerage. If we do not change course, disaster, perhaps catastrophe, is almost inevitable. As on the *Titanic*, the passengers in steerage will bear the brunt of the disaster, while the fraction of the passengers who can afford first-class tickets will sit in their half-empty lifeboats trying to stop their ears and cover their eyes.

There is a reason why interest in the *Titanic* has been revived; it's the perfect metaphor for our planet. On some level we know: we are on the *Titanic*. We just don't know we've been hit. The mass death aboard the *Titanic* occurred because of a collision with ice in a juxtaposition of carelessness and bad luck. Our disaster comes from our collision with a greenhouse—the greenhouse effect and global warming—a problem of such massive inertia that it will be extremely difficult to stop.

Unless we are effective in stopping it, excessive greenhouse warming will be an unqualified disaster for humanity, and in 50 years will bring the human race *in extremis*. The four horsemen of the apocalypse will ride, but the end of the world will not come. That would be too quick and merciful. As the temperature rises, the reassurances will resound that all will be well and that humanity will simply adjust to this supposedly inevitable change in climate. Unfortunately, they have no basis in either analysis, or history.

No accurate analysis can yet be done on the interaction of the atmosphere, the oceans, the biosphere, and the human race. The latter is poorly understood and the former are by nature unpredictable and powerful. No one knows what will really happen. It is a mad experiment, like setting your house on fire to see if the smoke detectors work. No historical precedent is useful either, since never before has the human race occupied the planet so completely, nor strained its resources so excessively. The only clue that history offers is that natural climate changes have usually inflicted great suffering and incited great bloodshed among humankind. The human condition is prone to misery even in times of stable climate and tends toward war even when no environmental stress is present. It strains belief to consider that the lot of the human race will improve as the world grows hotter, the winds faster and the Arctic ice more tenuous. However, this will not stop expert reassurance from becoming an industry.

People will shell out good money for reassurance. Companies will continue to set up institutes of global climate change reassurance, just as the tobacco industry has set up its own institutes and commissioned its own studies on the safety of cigarettes. Not surprisingly, these studies find no connection between cigarette smoke and cancer.

Since a climate change such as the one we are producing cannot be modeled accurately, reassurances concerning global warming are reminiscent of the rationalizations made by the officers aboard the ship, the *Californian*, who ignored the sight of repeated distress flares fired from the *Titanic*, which was at that moment only 10 miles away, stopped in the water, sinking, and surrounded by icebergs. The problem with the crew of the

Californian was that they lacked imagination. In the cosmos as we now understand it, lack of imagination can be fatal.

Perhaps on the *Californian* there existed a deeper problem: they may have actively suppressed their imagination. They were perhaps afraid to imagine that the rockets meant a ship out there was in distress in the ice-choked seas. This would require them to take action, to risk being wrong, to risk their own ship.

When the morning came, the *Californian* still took two hours to cross the 20 miles to the site of *Titanic*, but by then all that remained to do was to recover the dead. Greenhouse warming, it seems, is treated with the same detachment. Few corporations have bothered to examine seriously the titanic global collision approaching us. So, in great numbers, privately, publicly and corporately, we continue to load our atmosphere with greenhouse gases, adding daily to the problem. It would be too painful to stop, we are told, perhaps even dangerous to our economic health. The consequences will probably not be that terrible when they arrive, we are reassured. But this opinion flies in the face of virtually all of the evidence.

In the tropics, where most of the Earth's people live, often in profound poverty, often crowded onto small parcels of arable land that produce barely enough food to live on, any change of climate will bring terrible suffering and disruption. Global warming means spreading deserts and rising seas, which in turn means that the habitable regions of the Earth will shrink. If there were fewer human beings on the planet it would help, but there aren't "fewer." There are 6 billion of us, trending toward 11 billion, behaving like parasites, sucking the vitality from our host, choking the air with our wastes, unable or unwilling to control our populations, ignorant of how to contribute to the biosphere that supports our economies, feeling entitled to survive and prosper, but, instead, following a precise formula for creating a global disaster, perhaps a catastrophe.

Deforestation and its predictable second act, desertification (making desert-like), are the major activities of humanity at this moment. These low-tech ways of destroying the environment give the Third World its own share of the guilt, even as they point fingers at the developed world's contributions of CFCs, carbon dioxide, and industrialized fishing. In many tropical areas, as the rainfall patterns shift, deserts are being created where trees have been removed. These deserts are parched and stripped of every living green plant by the local people and their livestock, unwittingly accelerating Earth's decline as they struggle to survive.

Ultimately, the increased amount of rain that arrives after a drought may produce deserts too, rather than farmland. Because the heavy initial rains strip the already overgrazed hillsides of soil, and desperate refugees and their herds follow the rain wherever it goes, there is every likelihood that the land will be stripped of any new greenery faster than it can re-establish itself. Thus, we have the awful synergism of 6 billion humans—with many hundreds of millions already in distress from the shifting patterns of rainfall—adding to greenhouse warming by reducing plant growth and, hence, the amount of carbon dioxide removed from the atmosphere. There is already a net shrinkage of arable land in the Third World, and there is no upside to this.

As the oceans rise, many low-lying countries, such as Bangladesh, and island states are losing much of their arable land due to inundation by salt water. From the "back side" the deserts advance from drought and fire. What the fires leave standing, the inhabitants cut down for fuel; what the cattle leave, the goats eat; and what the goats leave, the people dig up and try to eat when they have nothing else left.

Ed Abbott was one of the noble and rugged characters who inspired me in my youth and early adulthood. He was a career forest ranger in Medford, where I grew up, and the father of my best friend. Ed was a sturdy man, conservative and forthright. He had defended and sustained the forests of America against fires and bad logging practices all his working life. He had seen friends die beside him on the fire lines as he fought back the flames, ate smoke, and swung his Pulaski through searing exhaustion in order to save the lives of trees. He'd manned lonely lookouts on the high ridges deep in the woods for weeks at a time. One time, alone on patrol, Ed had been bitten by a rattlesnake and had to ride horseback for 10 miles to get medical care—not as a straight shot down a road, but picking his way through the woods on uneven ground.

When I was a young man I would talk to Ed Abbott about the forestry issues of the day, particularly the clear-cutting versus selective cutting debate, an important issue in Medford where lumber was the major industry. He would insist that clear-cutting was fine, that the trees would regrow faster that way. This was the standard line at the U.S. Forest Service and Ed was very loyal to what he trusted to be enlightened forestry. As he was a man I respected, there was no reason for me to doubt him.

Many years later, long after I had grown to adulthood, I saw Ed again. We were relaxing and talking, enjoying the mountain scenery above Medford that we both loved so much, and sharing about old times, about how things had been when we'd both been much younger. Suddenly, a large barren expanse on a nearby mountain, many acres in size and nearly treeless, caught my eye and the memories came flooding back. I pointed to the clear area and said, "Ed, when I was in first grade I remember that mountain was clear-cut. It's 30 years later now and it's still bare up there."

He looked up and smiled, and with a twinkle in his eye he turned to me and replied, "That's just because you aren't looking close enough." Ed knew that clear-cutting hadn't worked there, but he didn't want to dwell on the mistakes of the past. Instead, he planted trees.

When Ed retired from the Forest Service after 30 years, he went around the world acting as a consultant to governments of countries trying to rebuild their forests. His experience in the Yemen, on the arid Arabian peninsula, had been a foretaste of some of the problems that we now face in remediating environmental problems. Ed had helped replant a region of forest in the mountains, only to have the local herdsmen graze their goats on the young seedlings. Remarkably, grazing goats are precisely why the primeval cedar forests of Lebanon were destroyed more than a thousand years ago, never to return.[2]

The government's response was brutal—the soldiers shot several goats and arrested the herdsmen. Then a meeting was called between the government officials and the local sheiks. Ed was asked to attend to explain the reforestation effort. In a bare concrete room, in a hard and alien land, Ed made his case to the sheiks through an interpreter, with the government officials, all armed, looking on. After this, the government once again laid down its harsh law to the sheiks. The sheiks stared at Ed Abbott and, as he watched, several of them drew their thumbs sharply across their throats, indicating in one eloquent gesture what they thought of his reforestation efforts, and what would happen to him if they found him out in the mountains alone.

This is but one illustration of the difficulties we face when we attempt to change long-standing economic practices in any part of the world to achieve preservation of the environment. This includes our own part of the world, wherever we may be. We may think it perfectly reasonable to ask that Yemeni herdsmen keep their goats from nibbling on tender seedlings; however, it is just as likely that we will be the ones who want to retaliate if the solutions imposed on us seem unfair or too draconian, or if they threaten our comfortable habits and cultural patterns. This imperative to change will cut all ways.

Ed is gone now, but the forests he nurtured and the trees he planted still grow all over the world as his legacy. I am grateful to him and his example.

It is proverbial that old men plant trees as an act of faith, precisely because they know they won't themselves live to sit under their shade.

Gerald Ford and Jimmy Carter, former US Presidents, *A Time to Heal*

In February of 1989 the Voyager spacecraft passed by the planet Neptune. Far different from the nearly featureless green orb of Uranus, Neptune is a deep-blue ball, striped with bands of color, and has its own great dark spot similar in proportion to the Great Spot on Jupiter. Who said the outer solar system would be a boring place? It was an amazing, dynamic, visually fascinating place and the grandest of the grand tours. Before it finally left our solar system to travel endlessly amid the stars, the Voyager sent back spectacular close-ups of Triton, Neptune's pink and blue Art Deco moon, showing evidence of cryo-volcanism—in which a cold slush of nitrogen and methane "lava" flows from vents across the landscape.

Our rather literary and astronomically-minded receptionist, Shelly, engaged in good-natured debate with me over whether the Voyager mission's omission of Pluto in favor of Titan was a tragedy or a triumph of good sense. Was Pluto really a planet? "Yes," she insisted. "After all, it has a moon and an atmosphere."

"Not a planet," I insisted. "Its mass is 1 percent of the Moon's. It is a ball of icy fluff, and should be reclassified as a comet. Or better yet, as a cast-off moon from Neptune." However, I observed with a chuckle, powerful vested interests within the planetary community would seek to preserve Pluto's planetary designation. If Pluto was downgraded to a comet, funding for probes to the last unexplored planet would lose their grand sound.[3]

Whatever the outcome of the continued designation of Pluto as a planet, it had no consequences for our own Strategic Defense Initiative research funding: it was going down. Ironically, as the grand tour of the solar system ended, so did the Cold War.

Vince DiPietro, Greg Molenaar and I published the "Cydonian Hypothesis" in 1990.[4, 5] We intended it as our final statement on the Mars problem until new images of Cydonia were taken and, at that time, no one knew when that would be. This of course, was one of the problems with Mars research. The missions then were few and far between.

In our published work we hypothesized that Mars had possessed an Earth-like climate and biology for most of its history, perhaps for as much as 4 billion years (we estimated that the Cydonia site was 500 million years old, based on a early model of crater dating by William Hartmann),[6] and that Mars, like Earth did later, may have evolved a humanoid-like intelligence. We based this hypothesis on the evidence of old water channels, the appearance of the Cydonia objects, and the principle of mediocrity—the principle that the Earth and its biology are "mediocre" rather than remarkable or exotic in the cosmos. We ended our article with the statement ". . . Mars once lived as the Earth lives now and it perished as surely as the Earth will perish if we are not better stewards of it." In our mind the fates of the two planets were linked.

The Cold War fell apart in 1989. The Russian occupation of Afghanistan

159

also ended, though, tragically, the carnage there did not. The violent struggle in Afghanistan continued unabated with new players. One of my neighbors, an Afghani, lamented that most of his family in Kabul had died since the Russians had left his homeland. He insisted the U.S. should send peacekeeping troops to end the bloodshed. I told him that, as regrettable as the situation was, intervention by the U.S. was impossible.

At Christmas, the Berlin Wall fell. I was at my family's home in Oregon for the holiday, and in a rare celebration of our German heritage, we Brandenburgs raised a champagne toast to the opening of the Brandenburg Gate. The "outbreak" of peace had a disastrous effect on my work in the defense area, but I comforted myself by remembering that this was the price of peace. I considered unemployment a more than even trade for the end of Communism and the lifting of the threat of nuclear extinction.

Inspired, I ordered and received a certified piece of the Berlin Wall as a memento and, during what I hoped was a brief hiatus, took a quick trip to Europe and lingered in the beautiful environs of Paris. I found that Europeans were both ecstatic and nervous about the collapse of the Berlin Wall, feelings I could easily identify with when I returned and found my job had truly ended.

Then Iraq invaded Kuwait on August 2, 1990, and, when they failed to release their stranglehold after repeated warnings, the U.S. and our allies, after a massive build-up, launched the armed liberation of Kuwait and an invasion of southern Iraq. The U.S., once again, was at war. I served on the domestic front of this war—at Lafayette Park in front of the White House in Washington, D.C., where I helped organize and run demonstrations to support the troops. The goal of our band of aging cheerleaders, which included Vietnam veterans and the relatives of soldiers fighting in the Gulf, was to get on the television news so the troops in the Gulf would know we were supporting them. We succeeded. It was at that point I realized the power of a few people, willing to make a public stand, to change the course of history. The troops did see us, and countless others like us, and they reported that the image of people in the street, waving flags on their behalf, filled them with hope and courage. As the war ended, several busloads of wounded, on their way from the airport to military hospitals in the city, stopped and thanked us.

Across the park the antiwar protesters, many of them veterans of the antiwar protests of Vietnam, stood and decried the carnage being done in the name of oil. At various points, groups from both sides would meet on the boundaries of our allotted regions, and with squads of police watching closely, we would debate the war. They would say that Kuwait used to be part of Iraq; I would say that Saddam Hussein liked to dip people in sulfuric acid. They would say that people were dying for oil, and I would tell them about people freezing to death in my home town during the oil embargo. Occasionally real dialogue would occur and I discovered that many of these people were quite rational and articulate and I ultimately became friendly with a number of them.

The issues were never as clear-cut as the stances each side was adopting publicly. Those opposing the Gulf War would admit that they were very uncomfortable supporting Saddam Hussein, and I would admit that our claim of defending democracy in the Gulf was something of a stretch. I told them that on a different issue I might be standing with them someday. I would also point out to them that Saddam Hussein was watching them on CNN every night and was counting on them to win the war for him.

In this war, like any other war, the environment was a victim. Saddam Hussein released a flood of oil into the Gulf, despoiling it. I asked our antiwar debaters how long they would be apologists for such despoiling of the environment, a question that would haunt me later. Many of our antiwar debating partners expressed great outrage at this shocking act by Saddam Hussein, and after this happened their numbers dwindled.

It got worse. On the eve of the ground war, Iraqi troops blew up and ignited every oil well in Kuwait, creating a cloud of black smoke that finally stretched around the world. Mercifully the war ended quickly after this, yet the loss of life was horrific. While the U.S. military refused to give a public estimate of the Iraqi dead, it was widely rumored in Washington, D.C. that nearly a third of a million people had died. The Gulf War was yet another reminder of how the industrial world's thirst for oil holds it perpetually hostage to the ever-shifting passions of oil-producing countries.

In 1992, on my whirlwind tour of France, Belgium, and England, I visited the JET (Joint European Torus) at Culham, near Oxford in the United Kingdom. The physicists there received me cordially and soon I was given a tour of the facilities. The Tokamak is a large, donut-shaped chamber 30 feet in diameter. The control room was very elegant: instead of racks of equipment, there were simply tables with desk-top computers, which allowed the whole experiment to be run from computer keyboards. The physicists were ecstatic that day because two days previously they had generated 200 kilowatts of pure D–He$_3$ fusion power in the JET—enough to keep a hundred homes warm in winter—without generating any neutrons or other radioactive products whatsoever. They had done this by superheating the helium-3 in the plasma by microwaves. It had taken five times that amount of microwave power to generate the fusion power, but still, I was highly impressed. Pure, clean power like starlight was possible. On that day I became convinced that not only was clean fusion power possible, but it presented the only real hope for human energy needs if our civilization is to continue.

Fusion power has been worked on for decades. It is hard to harness: that is one of its virtues. Fusion reactions require the Earth-based creation of conditions similar to the core of a star, and this is hard to do. Consequently, it is difficult to use fusion for weapons, especially if the fuels used are deuterium and helium-3.

The Gulf War may have been fought, in part, because of an effort to deter the development of fusion energy. Although the true motivations might never be known, it was widely speculated that Kuwait, with one of the lowest production costs per barrel of any OPEC nation, created the oil glut of the 1980s by pumping oil above its agreed quota, knowing that every other OPEC nation would have to follow suit to preserve its income as prices fell. In Iraq, Saddam Hussein's oil profits were threatened and he, in particular, became outraged with his southern neighbor. The reason Kuwait may have initiated this ultimately disastrous gamble was to drown energy research in the United

States in a sea of cheap oil. Eyeing the slow but steady progress in fusion developments in the '70s, when research budgets were growing beyond half a billion dollars a year, the Kuwaitis wanted to preserve the value of their one great national resource.

Just weeks before the Gulf War, the U.S. government declassified many of the technologies developed by Sandia Labs (where fusion experiments were being successfully carried out) to encourage their private development. This was done with great public fanfare, and press statements spoke about the new economic security and competitiveness that would be made available to the United States.[7] Perhaps the looming possibility of losing the market for oil altogether reinforced Kuwait's reluctance to back off. Whatever the motivation, Kuwait defied the OPEC guidelines and invited retaliation.[8] Despite the fact that it led to the country being almost destroyed, Kuwait did succeed in creating a glut in oil supplies. This reduced any sense of urgency to develop alternative energy sources, and the funding for fusion energy research and development vanished. This is a stern reminder of the lengths to which people and nations will go to protect both their oil supply and the supply of money that oil produces.

But it isn't just oil-producing countries that have the capacity to act irrationally. How can we hear and read about the number of species of animals and plants which become extinct and not be concerned that whatever is killing them isn't getting to us, too? How can we see frogs with extra limbs and not be concerned for our own children, for ourselves? What illness of mind has made us think that we are exceptions? ... that we are immune, exempt, untouchable? Do we really believe that we can carry on blithely doing what we have always done while the plants, animals and vulnerable humans of the world suffer and die? Is this how it was in Germany during the Second World War? Were people so brainwashed, numb, self-serving or in such spiraling self-deception that they just would not, could not, see? If this isn't denial, what is? What is it with us? Are we insane?

Yes. Our relationship with the Earth's environment is just as insane as Nazism or slavery. There are countless numbers of us who are totally unwilling to look at what the costs are—even to our own bodies and those of our children. We are unwilling to change anything, feeling fully justified, righteous and entitled. Are we really choosing our cars over our children? Our comfort over our health? Our convenience over the future of our species? In a certain sense, these are precisely the choices we are making, but we clearly don't see it that way. Nor is this our conscious intention. We send our children off to school with inhalers, convinced that we have done our best for them, unrepentant for any part we might have played in their disease because we don't even see the connection.

While the childhood cancer rates go up, we make contributions to the various cancer research organizations. Some organizations are interested in finding the *causes* for cancers;[9] however, for a significant part of the medical community, the unexamined and unexpressed strategy is to "cure" cancers that have already been environmentally induced. If we are to prevent the causes of cancer, we will have to live differently and find new incentives. There are few market-driven initiatives to stop presently acknowledged environmental causes of cancer. The best the marketplace is ever likely to do is sell us cures.

Sperm counts in the United States fell by about 1.5 percent per year between 1938 and 1990, and this may be a factor in increased infertility. In Europe, the drop in sperm counts is twice as steep. Shanna Swan of the California Department of Health Services and other experts blame the decline on persistent organic pollutants, including pesticides such as DDT and industrial chemicals like PCBs.[10] Yet again, the primary focus of research on infertility is on cures. As long as we can invent a process to overpower the problem of infertility—even if it costs tens of thousands of dollars—we are not as interested in the causes. We prefer to live with the illusion that these are problems that are unrelated to anything we humans might have caused.

As long as we can buy a cure, are we satisfied? Maybe, but that doesn't mean that the fundamental bankruptcy of our relationship with the environment won't catch up with us, too. The damage that is done to the environment will impact each of us in countless ways.

- It may be getting harder for our bodies to obtain adequate oxygen.
- We are more likely to suffer from infertility or sterility.
- Our children (and many adults) are more likely to develop asthma.

- If we have asthma, it is likely to grow worse.
- If we have heart disease, it's more likely to be fatal.
- We will be more vulnerable to cancers of all types.
- We will show some of the signs of aging earlier. Our skin will prematurely age and be more susceptible to cancers, and our eyes are likely to develop cataracts or develop them earlier in life.
- We are more likely to be attacked by viruses for which there are no known cures.
- With the proliferation of antibiotic use, there may be fewer effective medications for our illnesses.
- Our children are displaying secondary sexual characteristics earlier. Menstruation will onset earlier in childhood.
- Our personal economies will be less stable and retirement security less assured.
- Our investments will react to changes in the environment and climate.
- Our food may be more expensive and periodic shortages may be common.
- The food we eat is likely to be full of pesticides or antibiotics, and otherwise chemically or hormonally altered or contaminated with antibiotic-resistant *E. coli*, and reduced in nutritional value.
- Fish and other seafood may be unavailable due to contamination or destruction of habitats.
- Water will be more expensive and harder to obtain.
- Our tap water is more likely to be laced with chemicals, hormones, and fertilizers and, if we aren't already, we will be drinking bottled or filtered water.
- We are more likely to be involved in a fire, flood, hurricane, tornado, drought, debilitating heat wave or other freakish climatic disaster.
- Our charity will be sought over and over again to assist others who have been involved in disasters.
- Our own family may require such charitable assistance.
- Our taxes may increase to assist with the ever more frequent and destructive weather-related disasters. For example, our communities may be required to build structures to reduce flooding or coastal inundation.

- Our homeowner's insurance will be more expensive or coverage may be reduced.
- Many homeowners will be unable to procure insurance for their homes.
- Due to increased costs and risks, housing generally will be more expensive.
- Transportation will be more expensive.
- There may be periodic shortages or rationing of gasoline.
- Electricity will be more expensive.
- The birds, butterflies and amphibians in our neighborhoods will greatly diminish in numbers, if they haven't already.
- The habitats for wild animals will be reduced.
- The region in which we live may have economic upheavals brought on by fundamental changes in climatic patterns. As a result, regional economies may be depressed for long periods of time.
- Traditional methods of farming, ranching, and fruit growing may no longer be possible.

While all of the above is likely to come true, the insanity of our culture makes it more likely that we will read this list and mentally prepare to cope with the problems rather than look at what must be done to prevent them.

It is also possible for us to change. Cultures do change. The moral bankruptcy of Nazism was revealed and thwarted and modern German society is vastly more just and compassionate. The practice of slavery has been abandoned in most of the world, and where it still exists it is widely recognized as an immoral and unjust practice. Yet, in both the case of German Nazism and the American practice of slavery, it took major wars and the loss of countless lives to turn things around. It is possible for whole social systems to change, even those tied closely to economics (as was slavery, in particular), but can it be done peacefully? Can we stop the insanity of our environmental practices without the wars that stopped such cultural outrages in the past?

It may even be possible for humans to learn from the mistakes of the past. In our view it is not an unrelated coincidence that the current German government is one of the few national governments in which avowed environmentalists represent a large and vocal minority. As Germany has vowed to never again allow the kind of suffering that once occurred there, the Greens can be seen as a natural outgrowth of the concern for humanity acting as a voice of consciousness for the health and safety of the natural world. We acknowledge their courage and the value of their message.

Yet, for many of us, the mere scope of all the environmental problems creates such an overwhelming sense of impending disaster that we resort to disconnecting from all of it. There seem to be just too many problems to face, and with few clearly planned strategies to deal with the host of woes, it is difficult to decide where to begin.

Perhaps Gordon K. Durnil, author of *The Making of a Conservative Environmentalist*, said it best when he stated:

> Why do I think it important to set environmental priorities? Human nature is the primary reason. Let me give you an example of how an average person, already quite cynical about the ability of government to perform its most basic of duties, might react to the news of a new environmental problem. That average person might say, "Okay, I have heard enough about lead to agree that it is probably harmful. I want to protect my kids from exposure to lead. And I might believe that dioxins (whatever they are) and pesticides are bad for me, if you did not also tell me that movie house popcorn, eggs and bacon, smoking tobacco, coffee, product packaging, landfills and incinerators, meat, whole milk, hormones injected into cows, methane excreted from cows, radon, electromagnetic fields, ozone, forestry management, nuclear energy and carbon-generated energy, hot dogs, herbicides, automobiles, plastic, asbestos, vinyl, breast implants and Mexican food are all bad for me. It is just too much to worry about, so I will worry about none of it."
>
> Successful environmental protection depends upon public pressure. When the public hears one side say everything is bad and the other side says nothing is bad, the thought is mentally excused from their concerns. No pressure is exerted. It is easier to believe that nothing is bad, rather than to believe that everything is bad. So we need to set priorities. We need to attack the problems before they happen, and we need some consensus on where to start.[11]

Durnil is accurate when he says that we will have to set priorities in order to be effective in resolving our most pressing environmental problems. In order to do this, we will need criteria to establish those priorities. We believe that this can and must be done, but before we suggest solutions, let's understand this part of the problem a bit better.

The extraordinary disciplined practices of the scientific method—forming a hypothesis, experimentation, interpretation, reaching conclusions, defending and revising and, finally, seeing the acceptance of the theories which successfully pass through the gauntlet—are foundation stones of our culture, our entire way of thinking. This methodology certainly dominates the way we think about Earth sciences: oceanography, climatology, meteorology, geology, biology, forestry, ecology, etc. Discoveries of science are applied to all areas of life, but some discoveries, while the result of good science, may still be diametrically opposed to other findings. For instance, we may use good science to discover how to induce cows to produce more milk, using hormonal manipulation. We may also use good science to discover that hormones from treated cow's milk are causing premature sexual development in our children. While there are undoubtedly teams of scientists using good scientific practice making discoveries on both of these phenomena, they may well not be talking to each other, may not even share a common scientific language or specialty, are unlikely to be working for the same organizations or institutions, and may have very different goals and motivations. However, each group, each finding, can still represent the best "science" available even when working in diametric opposition to each other.

It is not possible to use a simple good science/bad science evaluation to resolve the opposition between these scientific discoveries. There is far more to the picture than that. Can you also see that there is an underlying issue of values here, too? And possible trade-offs? There are a number of values implied: high milk production from cows, the health of children, economic profits. In this example the trade-offs might be healthier children versus more economically produced milk or, possibly, as an absolute best-case scenario, affordable milk for children who wouldn't otherwise have any.

While science has played a role in solving many long-standing problems, its discoveries have also created or exacerbated many problems as well. On the one hand, science helps crop growth by developing fertilizers; on the other, it discovers that fertilizers damage the oxygen levels in the water of the Gulf of Mexico. It cuts both ways sometimes. There is nothing necessarily sinister in this. You might even argue that this is what human evolution looks like: two steps forward and one step back, or, in the case of fossil fuel use and nuclear power, one step forward and two steps back.

So here's the point. Scientific research does not serve any particular set of values exclusively, nor should it. As a result, science alone is not able to establish clear, unequivocal priorities for humanity to follow, especially for the kinds of immediately critical decisions, environmental and otherwise, that must be made today. At some point everyone, scientists, politicians, students, homemakers, businesspeople, and environmentalists—just everyday folks— will have to make those decisions because, as much as we hoped that scientists and technologists would solve the problems of the world, they need a clear agenda from us before they can do so.

The environmental issues we are facing are time critical. This is an emergency. There is so much at stake. We are truly facing life and death issues that will affect the entire population of the planet. Some of us, even tens of millions of us, possibly hundreds of millions, perhaps even all of us, will be dramatically affected. This is precisely why every voice counts. We must act quickly, especially as the scientific process doesn't necessarily provide completely solid answers fast enough for the circumstances we are in. But, often, excessively speedy research methods do not produce "good" science. Slow, careful, and methodical are the watchwords for producing good science through the traditional methods of research applied to environmental sciences. So, Earth's immediate need for environmental remediation may be thwarted by the slow progress of science. The model doesn't apply very well. It's not the time to do environmental impact studies when your house is burning down. First you put out the fire, then you figure out what went wrong.

There are other scientific models, however. There are models that have a sound grounding in scientific reality that do allow remedies, based on the best available information, to be applied quickly. Given the terrible consequences we may face if we wait too long, we propose to begin healing our environment, and to consider adapting other models for our purposes.

The first model we should consider is medicine. In the practice of medicine, lives may be lost if treatment decisions are not made rapidly. Medicine has developed the ability to routinely handle emergencies on a very rational basis. Every paramedic, emergency-room staff member or nurse in a clinic is trained in medical policies and practices that enable them to make quick decisions. The practice, developed by medics on the battlefield, is known as "triage."

Triage is a French word, originally used by French wool traders, meaning "to sort or select." In its current usage, it is the practice of categorizing patients according to a previously established set of criteria. Those categories may vary a great deal according to the circumstances, but the creation of a hierarchy based on the seriousness of the conditions is always a part of it. Life and death isn't always the issue. For example, in a busy pediatric clinic a triage nurse may prioritize patients over the phone in order to determine who would be best served by seeing the doctor immediately, who could wait until the following day (for a school physical, for example), who should probably have a day of rest at home, and who should go immediately to the emergency room. Under another set of circumstances, say at the site of an accident, the medical practitioner making triage decisions will assess the severity of the injuries to determine who can and can't be moved, who needs immediate surgery, who can tolerate a short wait but probably needs a transfusion of blood, and who needs just a stitch in the forehead or a finger set and can wait an extended time for treatment without this situation deteriorating. Sometimes, under the most desperate circumstances, for example on the battlefield where resources or medical personnel might be very limited, those performing triage will be assessing the conditions of patients to try and save as many lives as possible, putting the limited resources to use where they are most needed and will make the most difference.

The bottom-line value of medical triage is to save and assist as many people as possible using the resources that are available. We propose that the practice of triage be applied to environmental issues, and that a system of objective criteria be established to allow us to assess the level of urgency in order to be more strategic in how we tackle problems and use our resources.

Let's go back to the bewildering array of problems that Gordon Durnil talks about above. In an ideal world we might want to give each of these issues our attention. In the real world, it is impossible to juggle so many concerns at the level of urgency that is implied in the press releases. We very much need a rational system to help us. Where is our immediate attention required for our survival? What's the next highest level of urgency? Which issues are planetary in scope? Which issues are at the level of personal responsibility?

Durnil's list is far from complete. Obviously, there is a long list of concerns that aren't included—overpopulation, biodiversity, water, fertilizers, breast cancer, to name a few. The list goes on. However, a system of rankings could help us establish priorities, particularly if it were undertaken by a multi-disciplinary group including both medical professionals and scientists representing the various physical sciences, as well as representatives from the public. The rankings need not be binding; rather, they would provide guidelines to help us make the sound decisions that are so clearly needed. If we can rate movies based on content to help consumers decide what to see, surely we can rate the urgency of environmental issues.

There is another way of thinking about the environment that may help us sort this out. As a collection of societies and governments we are able to manage very nearly every aspect of a vast global economy. There is little in the economy that is unaccounted for. Those who steal and abuse money are viewed as criminals and a threat in nearly every society. The environment, the very basis of every economy, deserves the same level of management and concern. On the international level, for both the environment and business, precisely the same level of rules, foresight, responsibility and fairness should prevail.

Change, even positive change, has a way of seeming radical at first, no matter how reasonable it turns out to be. This is as true today as it was over a hundred years ago when a very new idea for saving lives was recommended. It simply didn't seem plausible at first. Not remotely. In the midst of heated battle in the Crimea, with wounded soldiers everywhere, an audacious report arrived suggesting that more soldiers were dying as a result of their medical treatment than as a result of their battle wounds. Such a blasphemous idea! How could one suggest such an ignoble fate for men who had fought so bravely? Nevertheless, there it was, an offending report proposing a long list of recommended reforms which were to be presented to Parliament and Queen Victoria as well.[12]

The conclusions that the report offered might have been easy to dismiss but for one factor: no one had ever seen anything like it before. It was an amazingly modern creation. In addition to a detailed, well-conceived text, there was a series of graphic drawings in a variety of colors that depicted, by size and shape, a body of numeric information that had been collected and interpreted with an unprecedented level of precision and sophistication. What mind had conceived of such a method for collecting statistical information on all manner of medical information and presenting it thus? It was impossible to argue with a mind capable of such methodical strategies.[13]

The report reflected a new set of disciplines and practices: the numerical analysis of meticulously charted records. Upon admission a soldier's condition and diagnosis were noted, as was his condition at the time of discharge or the reason for his death. Until this remarkable way of looking at the "behind the front" battle for life, it was assumed that rampant death of the wounded was to be expected. Yet this report—along with the statistics to back it up—suggested that simple improvements to the sanitary conditions in military and field infirmaries, like ensuring that laundry was done and that the floors were regularly washed, would help to save lives.

When the recommendations were enacted, a remarkable number of lives were saved. Far fewer wounded soldiers became statistics; instead, they left their infirmary beds and went home to join their families and live out their lives. While it took another 17 years for humans to outsmart microbes, it was not essential to confirm the theory of contagion in order to prevent deaths caused by it. By acting with the conscious, disciplined intention necessary to effect the changes suggested by this analysis, much suffering and many unnecessary deaths were avoided—even though it required inventing whole new disciplines. This is a remarkable admonition for our own age, which usually demands overwhelming proof of cause before taking action. Clearly, the time to take action is when you know how to make a difference, not when all the evidence is in.

The author of this revolutionary plan was Florence Nightingale, owner of a brilliant mind and a compassionate soul, master of medical statistics, inventor of both medical charting and pie charts, and founder of modern medical nursing practice. She saved the British army at Scutari and reduced the death rate from disease from 42 percent to 2 percent in just a few months during the Crimean War in 1855, forever transforming medical practice in the process.

There is something else you should know about Florence Nightingale. She had to fight to study mathematics. It was an education that would have been denied her had she not fought long emotional battles with her family, insisting that she be allowed to learn. A good education is still critically important—and not guaranteed to most of the world's population.

The education that is vested in you is a treasure, a gift. It is also a privilege and an investment that many people—generations of humans—have contributed to. Now humanity needs the dividends—the ones that you uniquely can generate. We are counting on you to save us.

CHAPTER 9

Lost and Gone Forever

They stand mute in museums—horrible, massive, with teeth the length of your hand. Once they ruled Earth in a reign that made modern human despots look like Quakers. Now, all that remains are their bones. Although still gargantuan and frightening to children as they leer from eyeless sockets, they are, nonetheless, endlessly, irrevocably dead.

The story of the extinction of the dinosaurs slowly revealed itself throughout the '70s, '80s and '90s in a manner that appalled the dinosaur experts. For, in spite of their best efforts, their territory, paleontology, had been infringed upon by the physicists and chemists. So, to their dismay, the first clue to what may have truly befallen the dinosaurs was the discovery of an increase in the concentration of iridium in the strata which marked the end of the Cretaceous Period—roughly the time when the dinosaurs disappeared. Worse for the traditionalists was the later discovery, in similarly-aged strata, of shocked quartz sites radiating outward from a presumed impact area (shocked quartz is produced only by the enormous pressures generated by volcanic explosions or by the kind of impacts associated with giant meteor strikes). These radiating rings indicated that this shocked quartz had probably come from an impact.

In 1990, Alan Hildebrand, with his fellow researchers, David Kring and Bill Boynton, found the smoking gun—the Chicxulub meteor impact crater, a 110-mile depression in the shallow waters off the Yucatan Peninsula, fitting more or less within the semi-circular area defined by the ocean area between the peninsula and the southeastern coast of Mexico. The scale and violence of the disaster was becoming apparent. The dinosaurs of North and Central

America had probably died almost instantly, as a result of the impact of a huge 4–7 mile meteor which crashed into the shallow ocean off the coast of Mexico. The resulting blast was so cataclysmic that its flashing heat triggered continent-wide firestorms which probably ignited the forests of most of North America. Eventually, the consequences of the impact wiped out not only the dinosaurs but also 90 percent of the creatures living in the sea, in a holocaust of almost unimaginable proportions. So the "end of the world" for the dinosaurs began with a fiery wall of death and destruction, a searing fire which filled the skies with an impenetrable layer of smoke and gases which completely obscured the sunlight. A chill quickly descended and then a deep, icy winter fell upon the Earth, locking the whole planet in months of frozen darkness. Nature had turned from nurturer to slayer. The dinosaurs, an advanced and colorful life form, full of promise and ferocity and with a budding intelligence, were destroyed in a geological instant, and the world was inherited by the scavengers—including the mammals from which we ourselves would eventually emerge.

As Earth's atmosphere was soothed by the cleansing rains, the huge amounts of carbon dioxide thrown into the atmosphere by Chicxulub, combined with the warmth of the still-dutiful Sun, and a quickened greenhouse effect, helped the Earth rapidly recover its warmth, and then some. Eventually, after many months, the atmosphere cleared and the Sun shone on Earth's surface again, inspiring what viable plant seeds remained to germinate and grow. In that first renewed warmth, a vanguard of young mammals were born, blessed to have no memory of the trauma, ready to take on life anew. For us, the progeny of these survivors, it is fortunate that memories too are not consciously passed down. However, in the absence of remembering, we must reason, and reason well—for the cosmos that we live in does not always give warnings. Dare we ignore those we are lucky enough to receive?

The Chicxulub mass extinction was not a singular event on our planet. It was only one of at least five great extinctions and many lesser ones that we know about. Of those, two seem to have been initiated by meteor strikes. However, 250 million years ago, 90 percent of animal life on Earth may have died from a huge release of carbon dioxide from the depths of the oceans. One of those five great mass extinctions, at the end of the Permian Period, was unlike the others in that the geological record shows no corresponding iridium layer for the Permian period (such as that later caused by the Chicxulub meteor strike)—which would suggest the extinction was not caused by a meteor. Instead, a group of researchers have recently theorized that several pieces of evidence point toward this extinction being caused by two colossal releases of carbon dioxide into the atmosphere, in what may have been a global "turning over" of the oceans—an event in which the ordinarily stratified layers of the oceans intermingle chaotically, with a resulting massive release of oceanic gases and great perturbations of the ocean's sediments. While it is unclear whether the carbon dioxide bubbled up from the ocean bottom like bubbles from a warm soda pop or was released in a sudden burst (as happened in Cameroon in the 1980s), the researchers suggest the amount of carbon dioxide released was so massive that it led to global warming on a scale which wiped out most life on Earth.[1, 2, 3]

This extinction scenario is both a timely reminder of the lethal potential of high levels of oceanic/atmospheric carbon dioxide and a warning about another of its chief characteristics: its instability when concentrated in solution with water. It also screams at us that today's carbon sink may be tomorrow's carbon source.

Humans are currently releasing massive levels of carbon dioxide, first into our atmosphere, and from there into our biosphere and oceans—and the rate of carbon release is increasing exponentially. With projections for 1998–2000, we now have a pretty clear record of human-created carbon dioxide increase since 1751. In just 250 years, by 2000, total carbon released back into the environment from its storage in the Earth's crust in the forms of gas, oil, coal, natural gas, and limestone, will be a staggering 284 billion metric tons.[4] As well as the amount of carbon released, it is the ever-increasing rate of release that is almost completely out of control.

Of the total amount, over 77 per cent (around 221 billion tons) will have been released back into the environment in the last 50 years—over 51 per cent (around 145 billion tons) in the past 25 years, and 25 per cent (71 billion tons) in the past 11 years. At those rates, we will double the amount of carbon dioxide released since 1751 by 2030, and triple it by 2050.[5, 6]

The primary way to slow these rates of increase would be to take on the massive restructuring of the world's economies, away from a fossil-fuel-based reliance. Unfortunately—even with the Kyoto Protocols—the huge growth of population in Asia and the world in general, and the Third World's legitimate desire to attain the same lifestyle favored in the West, make it highly unlikely (without major worldwide initiatives to reduce carbon emissions and to develop massive new sources of clean electricity generation) that humanity's yearly carbon dioxide production will be significantly reduced. Instead, the World Coal Institute projects the developing world to be a source of tremendously increased use of coal.[7]

We are also decidedly unsure about how much of the carbon dioxide goes into the different carbon sinks, and, particularly, how much can go there without "re-emerging" into the atmosphere in what is known scientifically as a "threshold event"—effectively, a scientific term for "the straw that broke the camel's back."[8] We suggest that by blindly pushing more and more carbon dioxide into the atmosphere and the rest of the biosystem, we could potentially be creating an atmospheric or oceanic catastrophe that is off the scale. There are indications we may already be pushing the limits. For example, in some places the rainforests have become net producers of carbon dioxide rather than net producers of oxygen. This could lead to out-of-control global warming—to temperature increases far greater than the gradual 1 or 2 °C often projected as the current rate for global warming over the next century. In fact, climate researcher Jeff Severinghaus pointed out at the United Nations Global Climate Talks in Buenos Aires that Earth previously has experienced temperature increases of up to 10 °C in as little as ten years.[9] The ocean/atmospheric system does not always change gradually in a "linear" fashion. Sometimes, when a particular threshold is exceeded, change is abrupt or "nonlinear."[10]

The best evidence indicates that we are presently putting about 2 billion more tons of carbon dioxide into the oceans every year than were entering the ocean before the use of fossil fuels.[11] When extra carbon dioxide goes into the ocean, a large amount of oxygen is leached out of the water at the same time and released into the atmosphere.[12] This could well have potentially disastrous consequences for ocean life (already the Gulf of Mexico suffers periodic bouts of anoxia—highly reduced oxygen levels—and the southwestern Indian Ocean is predicted to suffer similar problems). Also, and of greater concern, is the fact that until recently our atmosphere's oxygen levels have remained relatively stable, despite the infusion of additional oxygen from the ocean.[13] But now the atmospheric levels of oxygen are gradually dropping, even though more and more oxygen is being forced into the atmosphere by many billions of tons of oceanic carbon.[14] We are thus covering our oxygen deficit by making huge withdrawals from the oceanic oxygen bank. Our actual reduction in atmospheric oxygen must be much greater. Yes, we still look prosperous, with adequate oxygen and only moderate levels of increased atmospheric carbon dioxide, but we should be very concerned about the possibility that one day the oceans will return large amounts of the excess carbon dioxide stored in their depths to the atmosphere, in our hypothetical "Death by Soda Pop" scenario[15] (see Chapter 7).

In our rush to enjoy the fruits of a fossil fuel economy, it appears we may have been suckered into a game of environmental Ponzi, and, like all Ponzi schemes, unless quickly stopped and unraveled, it will likely collapse catastrophically. One of the worst possibilities is that the oceans could release massive quantities of carbon dioxide from their depths, while at the same time reclaiming oxygen previously displaced by that carbon dioxide. We don't know that this will occur, since we don't know the exact circumstances for such a release or how exactly such an event might be triggered, but one major group of scientists who analyzed the Permian extinction that occurred 250 million years ago suggested that the culprit was carbon dioxide releases triggered by tectonic activity or a chilling of the ocean surface waters by glaciers.[16] Another scenario which could release large quantities of carbon dioxide from the ocean depths could be that envisioned by the great geochemist Wallace Broecker. He suggests that a massive infusion of cold, fresh water into the North Atlantic, brought on by global warming, could stop or even reverse the Great Conveyor Belt, a massive flow of ocean water which brings warm air to the North Atlantic from the tropics and then pulls cold Arctic water into the depths of the Atlantic and returns it south.[17] If the Great Conveyor Belt stops, at best we will probably see a huge increase in atmospheric carbon dioxide levels for at least two centuries.[18, 19] However, there is no guarantee that the huge amounts of carbon presently stored in the oceans will continue to stay there.

While "Death by Soda Pop" is a worst-case scenario, unfortunately it cannot be completely ruled out, because no one knows for certain what the threshold conditions for carbon dioxide release from the oceans really are.[20] And, if significant parts of the large amounts of carbon dioxide currently stored there escape into the atmosphere, then we truly will have a catastrophe of epic proportions.

When we talk of "catastrophe," we are using the term as described by the mathematical model developed by the French mathematician René Thom in the 1960s. Catastrophe theory is a special branch of dynamic systems theory which studies and classifies phenomena characterized by sudden shifts in behavior arising from small changes in circumstances (like incrementally adding a bit more carbon dioxide to the atmosphere every day for year after year). One version of catastrophe theory describes the phenomenon of a sudden state change when there is a dramatic shift between two steady states (known as equilibria or fixed point attractors).[21] As applied to our hypothesis, in the present state vast amounts of carbon dioxide are stored in the ocean, but in the not-too-distant future, suddenly, when some unknown threshold is exceeded, much of the same carbon dioxide could be released into the atmosphere, potentially triggering a runaway greenhouse effect. Earth would become Mars by way of Venus, i.e., first becoming searing hot because of a hugely increased greenhouse effect and then becoming colder and colder as the heat gradually leaches off the planet's atmosphere and oceans—Death by Soda Pop.

However, even without a "Death by Soda Pop" scenario, we still face centuries of progressive global warming that will be brought on by the increased amount of carbon dioxide already in the atmosphere and by the additional carbon dioxide our continued use of fossil fuels has yet to release. Both sources will also push an already unstable planetary carbon cycle ever further out of balance. Beyond this, however, we are growing increasingly uncomfortable as we examine the evidence for possible catastrophe. For us, our discomfort started when we began to investigate where the missing carbon dioxide was going.

One of the great perplexities of global-warming analysis is that fossil fuel burning by humans is now putting over 6.3 billion metric tons of carbon dioxide into the atmosphere every year, but only slightly less than half of that remains there.[22] Of the remaining amount, current global modeling suggests that 2 billion metric tons are being absorbed by the Earth's oceans and 1.1–2.2 billion metric tons are being absorbed by the Earth's landmass.[23]

Given the dramatic difference between the absorbtion of 1.1 and 2.2 billion metric tons of carbon dioxide, it's obvious that global carbon dioxide modeling is still not an exact science. For instance, one recent study which tried to explain this discrepancy has suggested that much of the untracked carbon dioxide is being mysteriously absorbed on the North American continent.[24] This study suggests that the "missing sink" is to be found in the new growth of plants and in ground water. However, another study[25] says that this conclusion is impossible and that the North American carbon dioxide absorption figure is probably closer to 100 million metric tons.

We offer another proposal. We don't believe the "missing" carbon dioxide sink is solely on the North American landmass. We believe that much of it resides in the North Atlantic ocean and in the skies over the eastern U.S., Canada, and the North Atlantic. Perhaps, not so surprisingly, we are calling our hypothesis "The Missing Sink."[26]

To begin, consider these simple facts: Americans and Canadians produce about 25 per cent of the world's carbon dioxide, and yet, somehow, North America's carbon dioxide levels do not increase as much as would be expected. Thus the mystery of the missing sink.

Here are some of the clues to the puzzle. Carbon dioxide loves to combine with water to form carbonic acid. Its tendency to do this is so strong that, because of the absorption of carbon dioxide by rain, the standard measured pH levels of rainfall are usually accepted as being slightly acidic (about 5.6 or 5.7, with 7 being neutral).[27, 28] So even under normal conditions most of the Earth's rain is slightly acidic. Additionally, rainfall over the U.S. and Canada and up into the Arctic has increased dramatically since the beginning of the century; and if rainfall is up, net average atmospheric water vapor/carbonic acid is also way up.[29] So a large part of the missing sink may be found in the air, in the dramatically increased amounts of slightly acidic water aerosol or mist which must be suspended there to support the increased rainfall. An additional source of disguised atmospheric carbon is the huge amount of carbon dioxide that is neutralized on contact with atmospheric dust, which, most particularly in the case of desert dust, is highly alkaline. As a matter of fact, as much as 50 percent of the precipitation of carbon from the atmosphere comes in the form of particulate (dust) settling.[30] Once again, increased car use and fossil fuel burning produce increased atmospheric particulate levels, increased weekend rainfall and a larger airborne "sink" for the missing carbon dioxide.[31] One interesting indication of this is that acid rain readings from central China, an area with high levels of coal burning and other airborne carbon releases, showed highly alkaline pH levels of 7.9 in a recent student-conducted test—a bafflingly low result when one is expecting "acid rain," until you consider the high amounts of airborne dust created in central China from both the Gobi Desert and heavy industrialization. Ironically, high particulate levels can mask the presence of carbon dioxide.[32] So the atmosphere probably holds two additional major sources of carbon beyond those usually shown in carbon dioxide analyses.

Of course, eventually, much of the carbon starts to come out of its atmospheric suspension (by definition, over half of it) and is sent back to earth, some of it to be absorbed by plants and by natural geological weathering processes. Much of the rest then continues on and is deposited in the ocean, by several means: by settling of the particulates, by inflow of carbon-laden water from rivers and lakes, and, as the clouds and particulate travel over the ocean, by direct ocean surface carbon dioxide/water mixing, and by rainfall as well. The North Atlantic itself contains carbon in many forms, including, among others: a concentrate found in plankton and other ocean creatures and biota, carbonic acid, carbon dioxide, mineralized carbon forms, and carbon silicates. Most importantly, the cold, denser-than-water carbonated matter in all the North Atlantic's oceanic forms, will tend to sink to the bottom during the downward plunging of the Great Conveyor Belt, and be further concentrated on the ocean floor and the ocean-bottom sediments.

The ideal conditions for maximum absorption of carbon dioxide by water are, uniquely, in the North Atlantic. It has massive amounts of fresh water from ice or snow melt; vast biological activity, such as that of oceanic phytoplankton; temperatures just a few degrees above freezing which allow ideal carbon dioxide concentration in water (about 4 °C is best, and the North Atlantic's deep water is at about 3 °C); lots of precipitation and high turbidity of surface water for maximum contact between carbon dioxide and water and phytoplankton. So, we assert the "missing" North American sink is a combination of: the land's soil- and limestone-based absorption of carbon precipitates, the greater capacity of the atmosphere to hold carbon dioxide, and the concentrated carbon in the depths of the North Atlantic.

To give you an idea of the amount of carbon it can absorb, the North Atlantic is where six great flows of cold water (five of them fresh) join together. First are the large amounts of Arctic precipitation, which have increased 15–30 percent in the past 25 years.[33] Second is the huge volumes from small lakes, fens and streams covering northeastern Canada and feeding into Hudson Bay. Third is the flow from the Great Lakes into the St. Lawrence Seaway. Fourth is the hugely increased amount of North Atlantic sea ice and icebergs broken off glaciers. Fifth is the massive fresh water flow off the shrinking Greenland glacier.[34] These five sources of icy-cold northern waters add their combined flows to the world's greatest "river": the oceanic Great Conveyor Belt. The flow of the Great Conveyor Belt itself is almost incomprehensibly big—the volume of its flow is *twenty* times that of all the Earth's rivers together, and, as such, is capable of absorbing vast quantities of carbon and carbon dioxide.[35, 36]

Barring earthquakes, meteor strikes, or a vast heating up of the Northern Arctic and the consequent agitation of the deeper levels of the sub-polar oceans, tremendous amounts of carbon dioxide may tend to accumulate along the ocean floors. A large amount of this may be absorbed by calcites (mostly the exoskeletons of marine life) and by ocean minerals, but with a residue being retained in carbonic acid form, with the deep currents not penetrating deeply enough to produce much movement of these carbonic acid concentrations.[37] While it is generally assumed that this carbon will stay in stable suspension on the ocean floor for many tens to hundreds of years, in a summary article in *Science* magazine, Paul G. Falkowski says, "It is frequently assumed in biogeochemical models that this 'biological carbon pump' is in a steady state on time scales of decades to centuries. Such an assumption lends to the *unsubstantiated* [Falkowski's italics] conclusion that the biologically mediated net exchange of CO_2 between the ocean and the atmosphere is virtually nil."[38] We suggest that both the Permian extinction and the dynamics of the Cameroon killer lakes call this into serious question, and propose a far more unsettling alternative.

If the North Atlantic truly is a major part of "The Missing Sink" of North America, can you understand why our "Death by Soda Pop" hypothesis suddenly begins to look ominous? With eastern U.S. population centers on one side, and European population centers on the other, a massive carbon release (which, in an oceanic turnover, might also be accompanied by large amounts of highly toxic released hydrogen sulfide) could harm tens of millions of people, and, within a few months, cause Earth's temperatures to rise precipitously. Clearly, we must discover the validity of this prediction in order to avoid what, at worst, could become a re-creation of the Cameroon catastrophe on a planetary scale.

In 1988 suddenly almost all the cod, an ocean-floor-dwelling fish, disappeared from the North Atlantic. Scientists agree that this was not entirely due to over-fishing. Although the region has unquestionable been over-fished using destructive industrial fishing practices, one study suggests that of the 95 percent of cod that have vanished, 75 percent disappeared in a year in which cod fishing was banned.[39] Nonetheless, in three years the populations became so decimated that virtually the entire cod fishing industry came to a standstill—from which it has yet to recover. For instance, the fishing village of Port au Choix, Newfoundland, only ten years ago a thriving port, now stands virtually abandoned, because there are no more cod to catch.[40]

Were the cod the first to taste the Soda Pop and die? We pray not. However, after developing our own hypothesis for the missing North American sink, we noted that, in a study by Sarmeinto et al., the authors had to exclude the carbon dioxide readings for Nova Scotia's Sable Island station, right in the heart of what we are calling "The Missing Sink," when creating their model. These readings were thrown out, according to a senior National Oceanic and Atmospheric Administration official, because including them would have reduced the volume of carbon dioxide purportedly absorbed by the North American continent by 30 percent.[41] In other words, because of the unusually large volume of carbon dioxide being measured there, they assumed that the data from that station was incorrect. The tiny island, located 200 miles east of Halifax, is home to four scientists who operate the monitors which sniff for a variety of pollutants that spread northeastward from Boston, New York, and other coastal cities. We suggest that the Sable Island readings *were* accurate, pointing precisely to where part of the missing carbon dioxide sink is to be found—only a few miles from where the deepest part of the St. Lawrence Seaway channel empties off the Great Banks and into the deeper waters of the North Atlantic. An interesting bit of support for this comes from a recent interview in which Sable Island's research facility director, Gerry Forbes, complained that his staff members frequently had to wipe sooty crud from their instruments.[42]

We do not know how much carbon dioxide the oceans can hold before they begin releasing vast amounts back into the atmosphere, so we should not assume it is safe to sequester endless amounts of it there. It is safer to assume that both the terrestrial and oceanic carbon sinks are essentially full, and that at some time in the not-too-distant-future most of humanity's carbon emissions will directly enter the atmosphere and stay there. As a recent report by the American Geophysical Union concluded, "Because of the great vulnerability of socioeconomic and biological systems to discontinuities and high rates of climate change, studies of thresholds and nonlinearities in the climate and associated biogeochemical cycles are crucial areas of active research in the coming years."[43] While we are still pursuing the research we must err on the side of safety, since the physical and economic well-being of billions of humans is at risk.

It seems that we would rather find places to put excess carbon dioxide than give up burning fossil fuels. The United States, Japan, and Norway have joined in a research project designed to use the oceans as places to "dispose" of carbon dioxide on a long-term basis.[44] The plan is to pump liquid carbon dioxide in pipes to a depth of 3,000 feet, or greater, beneath the ocean surface, where they hope it with stay harmlessly for 600 years—unless it is suddenly burped back at us, refused by the ocean like the bounced check at the end of a Ponzi scheme.[45, 46]

It had become apparent that the two essentially important questions about Mars were, Did Mars live? and If so, for how long? Soon I was at a new job, one more directly involved with the space industry. I did not discuss Mars with my new colleagues until one morning, as I walked into work, a secretary asked me if I knew I had been on television the previous night. A rerun of a television program dealing with the "face" in Cydonia, in which I appeared for 60 seconds, had been shown. Cydonia had come back to haunt me again. As I explained my interest and involvement, my colleagues reacted with amusement. However, I became known as a Mars science expert in my professional circle.

Now working for one of the contractors that supported the U.S. Clementine mission to the Moon, I was honored to play a role in this adventure. But one aspect of this effort was particularly enlightening, made possible by the fact that astronomer and comet-watcher Gene Shoemaker was one of the other scientists on the mission. As we worked frantically to prepare the spacecraft and the mission support computer systems for the launch, we were treated to a curious briefing. Gene's wife, astronomer Fran Shoemaker, working with another astronomer named Levy, had been comet hunting one night and had found a beautiful specimen. This one had apparently passed too close to Jupiter and been flung into a crazy orbit up and out from the plane of the ecliptic where most planets orbit. We watched an increasingly spectacular event unfold, as Gene showed us slides of the comet's evolution over many days. It was a unique experience to see all of this "sky action" before virtually anyone else in the world. However, our astonishment deepened even more when Gene called yet another briefing a week later.

Comet Shoemaker-Levy 9, as it was later called in honor of its discoverers, was caught in the gravitational embrace of Jupiter. It was trapped in an orbit that would take it back to Jupiter. In fact, it was going to smash into Jupiter, in what would be the solar system's most violent show within human memory. A stunned silence filled the room. It meant we were going to witness a rare astronomical event in real time—as it happened, in one year. The human race was soon to get an object lesson in the wrathful face of nature.

As we worked on the Clementine and waited, the comet broke up into a stream of debris with numerous large fragments. At the time there was wide speculation that the expected massive impact would be decidedly mitigated because the comet was probably the usual "dirty snowball" and therefore its impact with Jupiter would be fairly benign. It was not until the day before the impact of the debris stream that I reached my own conclusion. It was not a comet, not in the usual sense, because it carried no water or other volatiles. Over time I had noted that the predicted water and other ice-forming gases had failed to show themselves as the comet debris train evolved. I preferred to describe it as an asteroid. So, the day before the first impact, I announced my own projection to my colleagues: Shoemaker-Levy was going to hit Jupiter like a freight train of dynamite.

Unfortunately the debris stream would approach Jupiter from the outer solar system, so the impact would occur on Jupiter's night side and not be immediately visible to us. Only when the rapid rotation of Jupiter carried the impact zone into view would we be able to see whether the fragments were feather pillows or rocks.

Like a long-awaited dramatic opening, the curtains parted and the debris fragments began to pile into mighty Jupiter right on cue. The results were both spectacular and mind-numbing. As they slowly rotated into view, we could see that the impact zones blasted into Jupiter's cloudy atmosphere were as big as the entire planet Earth. On and on, the bombardment went, for days, impact upon impact. The gleeful sense of watching a fireworks display initially felt by the Clementine team gradually disappeared in the face of the comet's ongoing raging, to be replaced with numb horror at the destruction being unleashed. Any single one of the multiple strikes would have devastated much of the Earth.

From space, Earth appears as a unity—a single, seamless circle of beauty set in a frame of darkened eternity. In the same way that we could observe the Shoemaker-Levy impacts on Jupiter from space, there are many things that we can see about Earth from the "outside." The ozone hole was first found from space. The fact that the Sahara used to be a vast jungle with flowing rivers was first seen from the Space Shuttle. Armed with special radar, it "saw" beneath the Saharan sands and revealed the river's courses etched in the bedrock. Like the dried rivers on Mars, we could actually see that a desert had been created in an area in which, until Earth's very recent geological history, vast waters had run free.

We are made responsible by what we behold; i.e., we are put in a position where we are able to respond, even if we choose not to. Having seen the havoc wrought by Chicxulub and Shoemaker-Levy, now we know that the cosmos offers no mercy to the ignorant. We are not just part of the greater cosmos; in an instant it can become part of us. We cannot pretend anymore that our cosmic danger is just an abstraction. There will be no shelter unless we make one. There will be no life unless we shelter it.

We are just beginning to take baby steps in protecting our planet from impacts. Although woefully underfunded, we've begun to search the skies for Near-Earth Objects (NEOs), which is the first step in protecting ourselves. However, at our current rate of progress it will take a century to take an inventory of 90 percent of what is out there, and that doesn't solve the problem of what we would do if we found something coming at us. We must do more.[47]

Soon after the Shoemaker-Levy strike on Jupiter and Clementine's triumphant visit to the Moon, our team received an invitation to a conference on planetary defense against comets and asteroids. We decided not to go because it was on the other side of the country and travel would be expensive. My boss, a very astute man and a pleasure to work for, remarked, humorously, that, with so many terrestrial problems, why should we worry about asteroids? I answered that, unlike crime, drugs, and other human problems, an asteroid on a collision course with Earth was one of the few problems we could solve. Saving the Earth and its biosphere from a comet or asteroid strike is an act of mutualism that humans, unlike any other lifeform, can perform. Apart from saving our species, it seems a gesture of great value, enough to begin repaying the Earth for what it provides us.

If you are aware of danger and do nothing about it, and the danger leads to tragedy, this is called negligence. It is a long-recognized principle in human society; if your ox tends to gore people, and you fail to pen it up, and it kills somebody, then according to Moses, your own life is forfeit.[48] I realized after Shoemaker-Levy that, if a large asteroid impacted Earth in the future, then the survivors would put on trial those government officials who should have known this was going to happen. Therefore, as Shoemaker-Levy riddled Jupiter, I found myself uncomfortably transfixed by this idea—with every good reason.

My words to the antiwar protesters came back to haunt me. I realized I was serving as an apologist for those who claimed that natural processes were causing both Earth's warming trend and the ozone hole that was now growing larger every Antarctic winter. I was violating my own sense of duty as a scientist in order to support those who shared my political views. But I argued with myself for a year. It took the killer heat wave that hit Chicago in 1995 to completely crystallize my views.

In a mere five days, beginning July 12, a total of 808 people died from heat-related causes in a modern North American city next to a cool, freshwater lake.[49] It gripped me with horror. It seemed unbelievable—like something that could never happen in my own country where every advantage is available. Many of the deaths were the result of the awful synergism of heat (temperatures reached 41 °C, with a suffocating humidity), poverty, and the societal chaos that afflicts the inner city. People, particularly the old, the young, and the infirm, died in their apartments because they were too afraid of crime to go out into the relatively cooler streets—or even, in a few cases, to open their windows. We could bewail the societal dysfunction all we wanted, but it was the heat that killed those people. It was then that I truly came to grips with the reality of climate change and its effects on human society. I realized I had been wrong.

188

As a political conservative, first as a Democrat, then as a Republican, I had looked on the environmental movement as merely an extension of the counterculture. I had written off many of its concerns as a rejection of the traditional American love of technological progress. Though I grew up in Oregon and loved nature, I did not feel any sense of kinship with the environmental movement, feeling that they had a political and cultural agenda that was diametrically opposed to mine, and that our shared concerns over the fate of tree frogs and the other helpless fellow passengers on Earth were somehow just an accident. I reacted to their warnings about coming global catastrophe with a scientific pose I rarely exercised in any other intellectual area—smug skepticism.

After the Chicago heat wave, I realized that the problem with a pose of scientific detachment and skepticism toward global environmental concerns is that it too easily translates into inaction. And Chicago had convinced me that inaction in certain environmental matters would kill people as surely as bullets.

Skepticism is one of the most abused attitudes of science. Unlike curiosity or desire for rigor, skepticism is often the fortified refuge of those who fear the truth. Skepticism can be extremely useful in quantifiable experiments that can be settled in a test tube. However, if the experimental test tube is the planet and one of the variables measured is the loss of human lives, then classical skepticism distances us from the reality of death and suffering. Instead, what must replace it is clear-eyed assessment of risk.

Assessment of risk in environmental matters involves identifying trends that are potentially disastrous, such as loss of ozone and global warming, and then considering the probability that specific human actions are contributing factors. We can then roughly calculate the cost, in billions of dollars or billions of lives, of the human behavior. It took Nobel Prize winners Sherry Rowland, Mario Molina, and Paul Crutzen's very careful analysis of ozone/CFC chemistry to describe accurately the ozone layer and to connect ozone depletion to human CFC production. However, it takes no particular genius to recognize that the cost of thinning ozone will be so high that even a low probability of human involvement justifies curtailing CFC usage. Thinking in terms of environmental risk allows us to make decisions even when information is not complete. In environmental matters, as in many other matters involving life and death, sometimes waiting for complete information means postponing action until it is too late. We need to call the Fire Department at the first sign of trouble so that there is still something left to save when they arrive.

As a conservative I don't like taking risks, not unless I have to. After Chicago, I quit relying on skepticism, and, instead, began to think in terms of risks. From this perspective, it's very logical that if we place our environment at increased risk by certain actions, we can change our actions to lower the risks. Conservation is a fundamental principle of conservatism. The conservative impulse, to lower risks, should apply to both the economy and the environment. After all, in a very real sense, they are the same thing. However, in order to recognize this, most of us would have to begin to think about things differently, to realize that some ways of thinking about what we see are more powerful than others.

When we think of a chair there are actually a number of ways that we might think about it. Most of us think of its function first, as a utilitarian object that we can sit on. However, there are lots of other ways to describe a chair which virtually ignore the fact that we can use it as a place to park our physique. We could see the chair as a three-dimensional design, like a sculpture. We might even view it in purely mathematical terms—as a series of dimensions in space—as we would a schematic drawing. We could view the chair in a historical context, seeing how it falls into the history of the development of "chairs." Or, we could view the chair as an economic entity, describing its production costs, value, and sale price. We could also describe the chair in terms of "matter," describing it as molecules, atoms, and subatomic particles.[50]

Can you see how all of these ways of viewing a chair (and no doubt many others as well) have validity? Although they may all describe exactly the same object, they do so in strikingly different ways, yet each is "accurate" given what we are calling here the "dimension" within which each description of

190

the chair occurs. So, we might think of the chair as existing in several dimensions: some physical, some conceptual, some economic, and so on.

It gets more interesting when you're able to make connections in the interrelationships between the dimensions on which the chair exists. For example, if the chair has a significant place in history, it might, as an antique, have a higher value than the ordinary economic value that would be warranted by basing it on its productions costs. If you altered its three-dimensional design by cutting off two of its legs, its utility as a place to sit would be severely altered. And if you began to alter its subatomic matter, say by dropping it in a nuclear reactor, suddenly all that would be left on the physical dimension would be the schematics and some subatomic particles. (However, sitting on it at this point could be very dangerous: you could end up creating a nuclear bum!) The utility of sitting on it would certainly be gone. The economics would also be altered (although how we would explain it on the insurance claim would be colorful). All that would be left would be a concept, a memory, and vapor. Yet notice that, although these ways of looking are describing different dimensions, they are all describing the same physical entity: the chair.

Another interesting thing you might want to notice about the chair and the numerous dimensions in which it exists is that most chairs are human inventions. So, while a chair's wood exists apart from human creation, nearly every other thing about it requires human participation and invention. In fact, as we invent new "sciences," the chair will probably acquire more dimensions. For example, the chair described in digitized data is fairly new, so the 3-D wire-frame electronic chair that can spin around on your computer screen represents a new dimension. Even stranger, if you are looking at a chair on one dimension, the others seem to be by and large irrelevant, almost as it they didn't exist.

Notice how complex a chair is when you begin to look at it from its multitude of dimensions, all of which co-exist. No one dimension can completely describe the chair. When you alter the chair on one dimension, the effects may take place on other dimensions as well.

It's helpful to understand the concept of the co-existence of dimensions, because one of the biggest worldwide problems facing us is a misunderstanding of this principle. Holding the co-existence of dimensions in awareness takes some practice. It requires that we be open to a multitude of significances, even those that seem to be paradoxical or that may not favor our particular viewpoint of the moment. The co-existence of dimensions requires us to respect that there are legitimately many ways of looking at things, that others may not be aware of the dimension we are looking at, and that we in turn may have little or no awareness of other dimensions. To have the most accurate perspective, we have to be open to looking at many dimensions.

We typically think of "the environment" and "the economy" as two different domains that are interrelated. Over on "the environment" side, we have trees and grass, oceans and land, people and animals, while "the economy" we see as the domain of money, value, production, commodities, and exchange. The relationship between the two is clear when we think of raw materials from the environment being processed into manufactured goods, or food production as a business that extracts and sells the resources of the environment. Often, though, we view these domains as in competition with each other—which we might express in terms of "we can't protect the environment at the expense of jobs," or, "the cost of environmental protection is a necessary cost of doing business." Notice that, by thinking of "the environment" and "the economy" as two distinct, competing domains, we are implying that humans have to favor one over the other. And so it often is in our world, where human alliances fall roughly into two camps: those who are committed to protect "the economy" and those who are protectors of "the environment."

As a result, we are frequently presented with issues where we have to choose between our environment and our economy—usually to the detriment of the loser. For example, we may have to choose between the jobs a new manufacturing plant would bring to our local area and the possible environmental damage generally accompanying that particular type of industry. Another choice might be whether to have an affordable car or a more environmentally-friendly and expensive car. We seem to face these decisions all the time in both our personal and professional lives. Our governments also see themselves forced to make these kinds of choices. So, how do you choose, when each decision seems to pit environmental and economic considerations against one another?

There are those who would fight to the death to protect their point of view as defenders of "the environment" or "the economy." Even the moderates in each camp have minimal tolerance for the values of the other. In effect, there is an ongoing, low-level, Cold War between the protectors of economic and environmental interests.

This is regrettable, particularly since it is most probably unnecessary. We put to you that this "cold war" is the result of a fundamental misunderstanding of the co-existence of dimensions. We suggest that you consider that the environment and the economy may not be two separate domains.

Here is what we see as a more accurate expression of the relationship between the environment and the economy. The environment is the economy. There is no real separation between the environment and the economy. They are the same thing.

The economy is simply a way of describing one dimension of the environment. "The environment" is the term that we use to describe the physical reality. Like the chair in the example above, "the environment" is tangible, but there are a multitude of ways in which you might dissect the concept of environment—for example, by viewing it through the lens of geology, ecology, meteorology, biology, or oceanography. Each of these scientific disciplines offers unique and extensive ways to look at a particular "dimension" of what we call "the environment." "The economy" is yet another term that we use to describe one particular dimension of "the environment." The economy is just another co-existing dimension, the dimension which describes the ways in which humans interact with the environment, and by which we trade one piece of it for another.[51] Each co-existing dimension has its own vocabulary, its own insights. Like the sciences mentioned above, each dimension has its own unique and methodical way of looking at and relating to the environment. Economists too have developed a whole vocabulary of terms to describe the environment—overheads, cost of goods, commodities, raw materials, land costs, equity, and money, etc.

Economies were created as a symbolic method of exchanging the underlying physical products of the environment—like trading shells or gold for grain. But even today, no matter what kind of currency is being exchanged, all the value and wealth of the world's economies lie inextricably in the physical stuff of the environment. The economy is an elaborate and abstract system which allows us to "model" or represent one dimension of the environment in a mathematical and procedural way. It is, nonetheless, only a dimension that we created by "agreeing upon" the system of representation. Over time, we've become so fluid in modeling the environment that it almost seems as if we have created a "separate world," disconnected from it. This world is so convincing that it even seems to compete for existence with the world it was designed to represent. But no matter how much we have bought into the illusion of separation, the brokers who have been selling the highly conceptual economic vehicle known as "derivatives" expect to be able to exchange the profits from these transactions for the goods of Earth, in the form of houses or cars, or other goods, and to do so at will, as if they indeed were equivalent. But what these brokers—and all of us—are missing is our "dirty deficit."

When we separate economy and environment and economic success becomes preeminent—when a company's stock drops 15 percent of its value because it only met, but didn't exceed, profit projections—then anything that hurts the bottom line must be avoided at all costs. The profits, too often, are pulled directly from the substance of the environment, where a corresponding physical "deficit" is created that no one intends to repay. Ironically, environmental deficits are often both generated and tolerated by those who would be absolutely up in arms if unexplained economic deficits—those belonging to another business—showed up on their own balance sheets. Thus we have a plethora of pro-business, anti-environment lobbying groups who distort and belittle environmental dangers in order to "protect" environmentally damaging practices masquerading as corporate profits. What if we protected the environment as zealously as we do profits?

What we are eager to distance from our awareness is that every time we ignore the "dirty deficit," every time we don't clean up a waste site or we denude a forest without ensuring regrowth, or spill toxic effluent into an estuary, we're killing another piece of the planet's ability to feed and sustain us economically in the future. In the name of a hidden deficit disguised as profit, of economy as if it were the only dimension of the environment that counted, we're peeling the planet—peeling it like an orange; tearing one piece of life after another from its surface, leaving another ragged wound which will fester, no longer able to feed or supply us. Huge strip mines in Indonesia tear holes in the rainforest; fishing trawlers scrape the ocean's ecosystem from its floors and strip-mine its aquatic life, taking everything indiscriminately, including the ability to sustain future life, future economy. In North America, peeling the planet takes a different form. We look clean and environmentally upright, but the U.S. and Canada produce one quarter of humanity's carbon dioxide, and, in the process, strip the atmosphere of its oxygen and fuel the gathering storms driven by an ever-increasing greenhouse effect.

Here's how Dr. Janet Yellen, Chair of the White House Council of Economic Advisers, describes the "dirty deficit."

> The Earth's surface appears to be warming from the accumulation of greenhouse gases from myriad sources worldwide. None of these emitters presently pays the cost to others of warming's adverse effects—a classic externality in the language of economists. As a result of these distorted incentives, disruption of the Earth's climate is likely to proceed at an excessive pace, and if left uncontrolled may pose substantial costs in terms of harm to commerce and the environment alike.[52]

The environment is the economy. In fact, we predict that the single most important determining factor for the world's economy over the next 50 years will be the environment.

The economy is an absolutely integral dimension of the environment, but it will exist only as long as humans do, and no longer. We must give up the illusion that it has a separate and competing existence. It doesn't. No environment, no economy. We created economies as a way of being more flexible in how we relate to the environment. In fact, we now depend upon the "economy" in exactly the same way that we depend upon the environment—for food, shelter, and every other material good we purchase or consume—but the economy does not have an independent existence. It is a co-existing dimension of environment. Environment and economy are just different ways of describing the same life-giving process, with economy being a way of measuring and accounting for the use of environment. However, when people cheat on their measurements and forget to account for and remediate damage to the vitality of Earth's biosphere as a result of their economic endeavors, then they are thieves of life, and their crime should be as abhorrent to us as any other form of theft.

On the other hand, when you can accept the idea of co-existing dimensions, it becomes clear that you aren't going to solve problems in the environment without fully respecting the economy as one of its faces. Environmentalists who would seek to pit themselves against "economic interests" are missing the point as well. For environment will not be saved in a world of people too poor and cold and hungry to care, and economy cannot prosper if the well from which it springs runs dry.

A new principle of accurate economics must accept that anything which has a negative impact on the environment will also have a negative impact on the economy, because, while the effects of environmental damage may not show up in the economy immediately, they will eventually. Environmental Ponzi will generate economic collapse as certainly as it produces environmental collapse.

We are just beginning to acknowledge the effect that global warming has on the world's economy. A recent study completed by the U.S. Federal Reserve Board suggests that " . . . El Niño appears to account for over 20 percent of commodity price inflation movements over the past several years. El Niño also has some explanatory power for world consumer price inflation and world economic activity, accounting for about 10 to 15 percent of movements in those variables."[53] El Niño is a climatic cycle that is now occuring more frequently as a result of global warming (nearly every other year). In fact, it is so closely associated with global warming that it could be argued that at this point El Niño is just one face of global warming.

Not surprisingly, one of the first industries to recognize the overwhelming threat of global warming to the world's economy was the insurance industry, because insuring against weather disasters is the one industry where accurate environmental modeling is a do-or-die economic proposition. If insurers guess wrong and set customers' insurance premiums too low, then the first major weather disaster will put them out of business. This actually happened to eight insurance companies in the wake of the twin American hurricane disasters of Hurricane Andrew in Florida and Hurricane Iniki in Hawaii in 1992, which combined to produce over $19.5 billion in insured losses.[54]

In response to these startling losses and company failures, the insurance industry was the first to step forward to discuss the real and potential economic loss that results from global-warming-induced climate change. At a 1993 conference entitled "Climate Change and the Insurance Industry," Frank Nutter, President of the Reinsurance Association, gave this prophetic assessment: "We in the insurance industry find that our economic interest is in fact very much intertwined with that of the environment and the climate. It is the threat of natural catastrophes that drives the demand for insurance products on property. It's also clear that climate change could bankrupt the industry." Insurance is essential in many industries to limit economic risk and liability. For example, all new construction requires insurance protection, as does home ownership that is financed by a mortgage. Without insurance coverage, many other industries would be at increased risk or unable to do business at all. As Eugene Lecompte, President of the U.S. National Committee on Property Insurance, put it: "So, the insurance sector could well be the first victim of climate change. As they are a victim, this might affect the stock market. That is the kind of scenario that many of us fear, that through climate change, that through the insurance sector, the whole financial sector will be affected by climate change."[55]

Back in 1993, when Nutter and Lecompte spoke of their concerns about global warming, the headlines shouted about the "Seventh 'Billion Dollar' Natural Catastrophe in Three Years."[56] For the 20-year period up until October 1987, the insurance industry had not suffered a single catastrophe costing more than a billion dollars in losses (in constant 1992 dollars), so seven weather catastrophes, each of which resulted in losses significantly greater than a billion dollars, in three years was a highly anomalous warning which galvanized the industry's attention.[57] Yet it was merely the tip of the iceberg. In a year-end estimate prepared by the world's largest reinsurance firm, Munich Re of Frankfurt, Germany, along with the Worldwatch Institute, the 1998 losses from weather-related disaster for the first 11 months of the year were put at 48 percent higher than the $60 billion record-setting losses in 1996. The 1998 losses were calculated at $89.7 billion, which, even adjusted for inflation, were more than the entire decade of the 1980s.

The toll in human suffering from these events was even more appalling. In 1998, over 300 million people, over one-twentieth of humanity, were displaced from their homes by weather disasters, and over 50,000 people died.[58] The economic losses have only just begun—we are experiencing the beginning of a trend that will unquestionably get worse because we have yet to start making a dent in the causes.[59]

When we understand that "the environment is the economy" we will be able to take actions that will help both. We have a habit of thinking that wants to fight the notion of environment as economy. To overcome this, the key question will be: Where are the opportunities available that accommodate the needs of both the environment and the economy? Or: If we don't protect the Earth, just which planet will we have our economy on? Clearly, if we really are going to create an alteration in the environment and protect the economy, we need to sit down immediately and work to get to the same side—to coordinate an effort to protect the environment and the economy together. If we can see the separation we create as the illusion it truly is, this will help. From this perspective, the way forward should be clear: we must guard the environment as securely as we guard a bank's vault, since our environment holds all the wealth we will ever have.

The Clementine was a revolutionary new way to build a deep-space probe. Its mission was to go to the Moon and then fly by an asteroid. It was "faster, cheaper, better," making use of the new miniature technologies developed during the days of Star Wars, with advanced on-board data processing and a streamlined management system for fabrication and mission control. The Clementine had another thing going for it, too: it was run by Naval Research Laboratory (NRL), with perhaps the finest team of space professionals assembled for 50 light-years. They were often young, enthusiastic, and absolutely dedicated to excellence.

As the build-up for launch to the Moon began, we watched with mounting excitement the arrival of the Mars Observer (MO), launched to Mars space ten months before. For Vince and me, this was our moment of truth. In the Cydonian Hypothesis we had made our case; now we hoped the pictures would be taken to prove us right or wrong. Vince, however, was deeply pessimistic.

"I don't think it's going to happen, John," he lamented. "I think something will happen so they won't have to take the images." It was a feeling that afflicted both of us, the feeling that Cydonia, if it was what it appeared, was too important for someone to just take a picture of it. It was the same old feeling that Cydonia was somehow forbidden.

I fought against the feeling. What could be simpler? Let's just take a picture and see what it shows, I argued. So, as the MO prepared to assume Mars orbit my excitement grew. The Clementine crew felt that 1993 would be a banner year for space science, with MO at Mars and Clementine at the Moon. Then, a disaster. MO turned off its transmitters, to secure them for the shock of the rocket burn, and was never heard from again. It vanished. Mars' space had swallowed her. Vince and I, despite even his pessimism, were shocked. The utter totality of the MO's disappearance, the fact that JPL appeared completely helpless, seemed to us, deep down, to confirm that Cydonia was, indeed, somehow forbidden.

I tried to overcome my edginess by concentrating on my work, supporting Clementine mission planning and finishing some long-due reports for NRL on a previous Earth orbit mission. As I sat at NRL working, a senior member of the Clementine team came to the door of the office.

"What are you doing here, Brandenburg?" he demanded.

"I'm finishing the last report," I said in puzzlement.

"Don't just sit there. We're going to Mars, and you're the Mars expert! So, get up! We need you down at mission planning."

Stunned, I stood up and followed him. At the mission planning offices there was an organized frenzy. NASA had requested a plan for a Mars mission, to be launched in one year, to replace the MO. I pitched right in. Suddenly all my accumulated study of Mars—all the papers, the books, the discussions, the conferences—paid off. We put together an excellent plan for a Mars mission, using a modified Clementine spacecraft. For about a week it looked like we would go. However, Mars is the most political piece of real estate in the solar system (after Earth, of course). It was jealously guarded red stone. JPL, aided by its Congressional backers, was fighting desperately behind the scenes to stop any Clementine mission to Mars. Suddenly, when the decision was assigned to committee, a committee composed of mostly JPL and ex-JPL people, our hearts sank. We were not even going to be allowed to present our own case, even that was going to be done by JPL. The outcome was as ludicrous as it was predictable: the committee decided to build a clone of the MO and re-launch it to Mars. It was felt that Clementine did not address the proper "questions" at Mars, and that its sensors were "inadequate," especially its high-resolution camera, the very one we had hoped to use to get images of Cydonia at better than Viking resolution.

So close, but still so far. However, JPL was determined that nothing it did not control would come near Mars. Mars exploration is a zero-sum game. In retrospect, their attitude was understandable. Any loss, especially after MO, would have been terrible, resulting in real harm to JPL's prestige, hundreds of layoffs, and ended careers. So, we of Clementine team refocused on our main mission with increased determination, and settled for triumph at the Moon instead.

Our mission to the Moon was spectacularly successful. We mapped the surface to then unheard of levels of high resolution and spectral precision, proving the worth of the "faster, cheaper, better" philosophy. We also made several major discoveries, the most important being billions of tons of frozen water at the Moon's poles, a discovery that was confirmed and expanded by a later probe, the Lunar Prospector.

We tried for the asteroid too, but, unfortunately the computer, the miracle that made the Clementine spacecraft so capable, crashed, just like a desktop computer, and, bypassing three separate safety functions, sent a bad command to the spacecraft's thrusters, spinning it up like a top and draining its attitude-control fuel. We managed to recover the mission to a great degree and start spinning it down, but the asteroid's window had already closed. The Moon's gravity finally flung the spacecraft out to orbit between Earth and Mars, where it was found, out of fuel, but in otherwise perfect working order, a year later—no longer lost, but gone forever.

CHAPTER 10

Dead Mars

I n 1995, I was asked to serve on a NASA advisory panel concerning space transportation. Our job was to aggressively reduce risks. This was a great honor and I enjoyed the experience immensely. Also about that time, Vince DiPietro had connected with Dan Goldin, the head of NASA, at a Goddard Space Flight Center get together. He'd handed him data on the microfossils in the CI (carbonaceous chondrite) meteorites. Goldin was immediately fascinated and said that he wanted to meet further with Vince, but an appointment proved difficult to arrange. I also encountered Goldin at NASA advisory panel meetings in early May of 1996, and, since my article on the CIs had been accepted for publication in a journal, I gave him a copy. I told him we could prove life had existed on early Mars. He appeared transfixed.

It had been difficult getting the CI article published. I'd debated with the publication's referees (some of whom were from labs at the Lunar and Planetary Institute in Houston) for a year about various aspects of the paper, carefully addressing each objection as it was raised. Finally, the referees raised no new objections and my paper was accepted for publication.[1] When I got the good news, I called Bartholomew Nagy and told him that the CIs were from Mars and that this meant that he, as the one who had discovered fossilized microbial life aboard the CIs long before it had been identified on the SNCs, had really discovered life on Mars.[2] He sounded pleased in a low-key sort of way. I sent him the article, but when I called back three months later to hear his reaction, I was saddened to hear that he had died a month before. At least he knew, to the extent I was able to communicate the news, that he had been vindicated.

Publication of the findings on the Martian origins of the CIs did not mean general acceptance of the idea that they were from Mars, of course.[3] One scientist was so outraged by my article that he yelled at me over the phone, saying that what I had done was "not science!" I encountered that special brand of hostility reserved for those who threaten to shatter the paradigm of human aloneness. However, the bitter turned sweet when, that August, the whole world was turned upside down by NASA's announcement that microfossils had been found in the first old meteorite from Mars—ALH84001. We'd finally done it! We'd made a difference. By offering up the CIs as Martian, we'd lowered the "risk" of finding life on Mars.

On the day of the Mars life announcement, Vince and I sat in the briefing room, watching history unfold. Dan Goldin stood tall in his cowboy boots, making the announcement concerning ALH84001, while a panel of scientists sat behind him. As Goldin placed the prestige of NASA behind this amazing discovery, I viewed this decision as the act of an intelligent gambler and an inspired leader. NASA's budget was being cut severely to help balance the national budget, and soon NASA would not have enough money to mount any major new initiatives—and the agency needs new initiatives to survive. If there was life on Mars, NASA now had a mission. Goldin was like a king, besieged in his city, who boldly sallies forth one night from behind the safety of the walls with his army, and catches the enemy force encamped against him asleep in its tents. After the main press briefing, Goldin stepped from the podium and wordlessly shook my hand. I was astounded and gratified. With the CI meteorites in his back pocket, perhaps he had known that the ALH84001 discovery would eventually be confirmed. In the meantime the Houston group from NASA who had made the finding of Martian life are now being chased by the mob of paradigm defenders.

Carl Sagan was terminally ill, but he lived to see the NASA announcement. Sadly, he died soon after, but he could be satisfied that his work helped to make that announcement possible. Despite his public hostility to aspects of the Cydonian inquiry, his quiet, behind-the-scenes interest in the work of the independent Mars investigation did much to encourage us.

Ironically, Carl Sagan made his reputation as a scientist by looking at the greenhouse effect. In its worst aspect, the greenhouse effect rules the planetary conditions on Venus today. There is some evidence that Venus, like Mars and Earth, may have had oceans early in its history. However, as our sun grew warmer, Venus was thrust into a runaway greenhouse effect, which occurs when the heat build-up of the greenhouse effect feeds back on itself. If there was an ocean on Venus, initial levels of greenhouse warming would have significantly raised the temperatures of the Venusian ocean. Since warm water holds less carbon dioxide than cold water, this warming would then have released more carbon dioxide from solution into the atmosphere, adding to the greenhouse effect and making temperatures rise even more. This would then have forced the oceans to release more water vapor, which is a greenhouse gas a hundred times more potent than carbon dioxide, thus vastly accelerating a cycle of compounding deadly effects.[4]

Venus may even have had microbial life before its temperatures were raised precipitously. We can predict microbial life on Venus because both Earth and Mars appear to have had life early in their histories—so early, in fact, that it is questionable whether it would have been possible for life to have spontaneously developed from the primordial soup on the individual planets. One theory that would refute the idea that life originated on Earth says that Earth, Mars, and Venus may all have been seeded with life from space. This theory has been reinforced by the discovery of microfossils in the Murchison meteorite,[5] a CM, the second category of carbonaceous chondrite meteors, not a CI. Reaching Earth in a fireball that exploded and sent hundreds of stones raining down over Victoria, Australia, in 1969, the Murchison probably came from the asteroid belt, but certainly not from Mars. So, with the recurrence of microfossils in yet another meteorite, suddenly it's looking like microbial life thrives widely in our solar system.

One of the earliest predictors of interplanetary transfer of life was the scientist Svante August Arrhenius, who won the 1903 Nobel prize for chemistry for his explanation of electrolysis. This "living cosmos" hypothesis was called "panspermia." Arrhenius thought the cosmos was crawling with life that was spread from planet to planet (and even between stars) by spores which were carried on the pressure of radiation. As unlikely as this theory seemed for many decades, the evidence is beginning to mount that Arrhenius' genius was greater than we possibly imagined, and that, while the delivery packets are more likely to be meteorites, life, in fact, probably does travel from planet to planet and maybe even from star system to star system. Remarkably, Arrhenius was the author of another powerful and sobering idea—that carbon dioxide could trap heat on a planet's surface and lead to a greenhouse effect. His seminal paper on the greenhouse effect first advanced the concept that by burning fossil fuels, and thus injecting large amounts of carbon dioxide into the atmosphere, humanity was changing the thermal balance of the Earth and producing greenhouse warming.

Carl Sagan was the first person to note that Mars' atmosphere could exist in two stable states. In one state, it was cold, with a thin carbon dioxide atmosphere, half of which which was concentrated in the form of dry ice at the Martian poles. There was no greenhouse effect because the atmosphere was so thin. In the other, there was an atmosphere of roughly the same pressure as Earth's, composed of carbon dioxide. The carbon dioxide trapped so much heat that the temperatures were like Earth's and liquid water could exist and move around as it did on Earth, with water vapor adding to the greenhouse effect. (Fortuitously, as this book goes to press, the scientific case for a liquid water state on Mars has just been powerfully supported by the announcements of the Mars Orbital Laser Altimeter team, which found strong evidence for a vast paleo-ocean on the Northern Plains of Mars.)[6] There was thus a distinct separation between these two states, and while Mars was in either one of them—frozen wasteland or in full greenhouse phase—it was fairly stable. But with one major event, like a large asteroid impact, Mars' climate would suddenly destabilize. Since huge flooding and resultant channeling seem to have occurred during the Hesperian Age on Mars (as can be seen from Viking images taken of Chryse Basin), it appears that Mars had been in its full greenhouse phase just before the Lyot impact and had then tipped over into the second stable state: the frozen state it is in now.[7] However, in the process, Mars had lost most of its carbon dioxide atmosphere and no longer had enough carbon dioxide to restart the greenhouse effect.[8] Mars was to remain forever frozen.

The Mars Global Surveyor (MGS)—NASA/JPL's first chance for new aerial images of Mars in over 20 years—arrived at Mars in early September 1997 and was to begin its mapping mission in March 1998. Unfortunately, during initial aerobraking, a solar panel came loose, so a more conservative plan was initiated and the mapping mission was rescheduled for 1999. In spite of that problem, the onboard camera immediately began taking pictures, some of them stunning.

After an enormous amount of maneuvering with and by NASA in private, and a torrent of public pressure, an announcement was made that new images of the features at Cydonia would be taken with the three opportunities occurring on April 5th, 14th, and 21st. Because of the controversy involved, two JPL officials promised independent Mars scientists that no official statements would be made concerning the images when they were released.[9] NASA insisted that Michael Malin, the private contractor charged with operating the MGS camera, take the pictures (despite his objections), but, like the life experiments on the Viking, the re-imaging of Cydonia was pronounced a fool's errand by many scientists who had never seriously looked at the data. When the image of the "face" was finally taken, it was under very awkward conditions, as different as it was possible to imagine from those under which the first two images had been taken in 1976. Nonetheless, it was new data, and the hard-won opportunity to obtain it was treasured by those who had waited for over two decades for even one new image.

As the first new images from Cydonia in 22 years were being publicly released, Vince DiPietro and I were in New Hampshire, working on making large balls of mid-air plasma, attempting to recreate on a larger scale in an industrial microwave oven our earlier success at making the first man-made ball lightning. It had been a long, exhausting day with little experimental progress. We returned to our hotel late and heard on the news that the new image of the face had been taken and had already been pronounced as just an eroded mesa—as Malin and other detractors had been saying for over two decades. Like a nightmarish replay of the rush to scientific judgment that had been made in July 1976 when the feature had first been discovered, an unscientific rush to instant summary judgment had overwhelmed any possibility for careful analysis of the new data.

Vince and I tried to download the image from the NASA Internet site to a laptop with little success. We walked to a nearby restaurant and ate, talked, had a few beers, then had a few more. We were both miserable. No one likes to be wrong. We went back to the hotel and saw the image of what we had worked for for so long on television for a few seconds and were astonished at how different it looked. As wretched as we felt, we both knew that in science you put forward your best hypothesis and then seek to prove or disprove it. We had waited literally years for this moment.

After returning to his room, Vince received a call from a New York talk show host, cackling like a harpy, who derided him for investigating the "face." "Don't you feel like a fool now?" the man exulted. Vince hung up on him and all his listeners, thus cheating him of his sadistic romp. I finally was able to download the image and really look at it for the first time. It looked like a pile of rubble. Alone with my thoughts, I had to consider the possibility that all I had done, all the ridicule I had endured, had been a waste. Ultimately, I refused to regret any of it. I had begun the exploration of Mars in Cydonia, and the inquiry had led me to Amazonia. The pursuit of the truth had taken me on an amazing path of discovery, leading all the way back to a tremendously increased new level of insight and concern for my own planet.

Vince returned to my room. We watched CNN while a senior MGS mission scientist was being interviewed. He was one of the two MGS scientists who a week earlier had promised not to comment on the new MGS images.[10] He looked drawn and uncomfortable. Although he was saying how it had been just a mesa all along, he came across as scolding and dour, not the relaxed and triumphant person I expected. I turned to Vince and asked, "Where is this man's joy? He looks like he just saw a ghost."

"I know," agreed Vince, as puzzled as I was. "Why isn't he happy?"

We spent a moment puzzling, then our conversation turned to the plasma experiments for the next day. Vince went off to his room and I went to bed, but couldn't sleep. I kept wondering why that MGS scientist had looked so miserable. I got up to look at the new image on the screen of the laptop again. Was there something there that had spoiled his feelings of triumph? I stared at the image very carefully, transfixed.

The giant asteroid tumbled toward Mars without remorse. The planet it was drawn to by gravitational force had no technology to anticipate its arrival and enable it to strike back. There would be no desperate salvo of nuclear weapons to deflect or shatter the asteriod as it came in. It arrived from the night side, the side away from the Sun. If asteroids had minds, it would have been afraid; if it had a heart, it would have wept; but it was lifeless stone. As the asteroid approached, it tore through the planet's atmosphere, showing a terrible brilliance in its passage, plunging down into the rock at the boundary of land and sea. From a distance the brilliant expanding fireball would have looked like a rising sun. The light was so intense, it ignited every piece of organic matter it shone upon. The shockwave that broke free from its surface created a vapor cloud in a dome around it, blocking the light for a moment before dissipating as the standing wave rolled out across the dark landscape. In the ocean, a sledgehammer wave of pressure swept the ocean clean of any life. Above, the vapor cloud's searing light reappeared as a fireball. Then a cloud of blue-white debris broke through the top of the atmosphere and formed a pillar reaching through the sky, as the shockwave slammed outward, flattening whatever vegetation Mars may have had, sucking everything back into the vacuum the blast had created as it ripped a huge hole in the atmosphere, letting frigid space pour through to finish the destruction. A shower of white-hot debris fell behind the shock, racing around the planet and igniting everything left. The sky was now a burning oven. The shockwave, focused in the southern hemisphere at some unlucky point south of Amazonis, was then reflected back to its point of origin with a force that would have been deadly had anything been left alive.

So Mars first baked and then froze. Its ocean was converted to ice; its carbon dioxide atmosphere dropped, frozen like broken glass, to eventually settle on the Martian poles.[11] Its oxygen atmosphere recombined with the soil and the organic matter left exposed, and the ultraviolet rays of the sun pounded relentlessly downward to finish off whatever was left alive on the surface. Even water broke apart under this bombardment and fled away into space. Mars rusted, its surface oxidizing to red, its hydrogen lost forever between the planets.

So Mars took its place among the dead and barren planets of the cosmos, and, as the dust storms blew, a lonely mesa that looked like a face stared out into space with lifeless eyes. The remnants of the great northern ocean of Mars froze, and then slowly retreated—both north to the polar cap and down into the planet, leaving the plains of Amazonis a frozen desert beneath the unblinking stars. The underside of the miles-thick layer of ice melted, forcing the water underground, imprisoning the frozen carbon dioxide (dry ice) mixed with it in the form of vast underground carbonate deposits, which displaced silica. The silica freed from the rocks formed a huge sand sea around the polar cap, larger than the Sahara. A sea of dunes; vast, beyond any on Earth.

Suddenly, staring at this ancient eroded form on my computer screen, I felt a familiar sense of dread. Had the asteroid strike at Lyot been witnessed by conscious beings? What if this had happened on Earth? It was a chilling thought but a profound reminder: planets live and planets die. Some die by random acts of cosmic violence and some, perhaps, by destructive forces unconsciously unleashed at the hands of their own inhabitants.

I had been trying to model Mars' atmospheric composition before and after its fall when I had a shocking relization about our Earth. I imagined that before the catastrophe, Mars' atmosphere had been similar to Earth's but with carbon dioxide replacing nitrogen as its major constituent (about two thirds), and, beyond that, water vapor and oxygen (about one fifth) and nitrogen (probably one tenth). To get an optimum greenhouse effect, you want carbon dioxide at about Earth's atmospheric pressure so that water can evaporate at normal temperatures and play its part in the process. I had imagined a green fertile Mars before the fall, with an ozone layer, or, perhaps, as recently suggested, a layer of frozen carbon dioxide, but, in either case, it may have had abundant life, like Earth does now.[12] Then came the asteroid and the endless winter, with the carbon dioxide and water vapor frozen out of the atmosphere, ensuring that the greenhouse was forever shattered. All the life, except for some faint microbial shadow of a biosphere, died, and a ghost of an atmosphere, some nitrogen and the oxygen, was left to slowly decay. The nitrogen would be fairly durable and inert and would survive, but oxygen is reactive and so would combine with the soil to make the reddish color of Mars we see today. But how fast would the oxygen atmosphere disappear? I found myself continually puzzling over this question during the winter leading up to the MGS imaging of Cydonia. I wondered why I kept returning mentally to this problem; it seemed unimportant. I had heard the estimate that it would only take a million years for Earth's atmosphere to decay if all life died, a blink of an eye in geologic time, where a billion years is considered but a moment. Applying this estimate to Mars' oxygen meant it would lose one part per million per year. I immediately thought of the only planet I knew which had an oxygen atmosphere. How long was its cycle time and what process governed it, I wondered? Then Stephen Corrick mentioned to me the fact that in some Japanese cities oxygen levels are so low at times that it feels difficult to breathe, while my co-author, Monica Rix Paxson, was alarmed by the rising carbon dioxide levels with its myriad implications, some quite unexpected. We also knew that ocean oxygen levels had to drop as they absorbed more carbon dioxide. As all these ideas circulated in my head, I suddenly realized that Earth must be crossing a fateful threshold.

Earth's oxygen level must be dropping, I thought, if not now, then in the near future. I knew that the Amazon rainforest produced a large percentage of our oxygen and that a significant portion of the world's rainforest has been slashed and burned. I checked the rate of atmospheric carbon dioxide increase, which is now approximately 1.5 parts per million per year and accelerating geometrically. Then I considered the effect of geologic processes, the reaction of carbonic acid with rock to form sand and carbonate, and volcanic emission of carbon dioxide from buried carbonate. About this time, Monica came to the realization that all the global climate modeling was, at one level, unnecessary to see that all sinks were at effective capacity. She called her discovery "All Sinks Plus." No matter how much carbon the biosphere processed or absorbed, for every single year back to 1750 the sum capacity of all sinks had been exceeded, with the overflow adding incrementally more carbon dioxide to the atmosphere. Assuming the rate of carbonate formation and volcanic exhalation pretty much balanced one another out, this left the rate at which the biosphere was making carbon dioxide by burning oxygen and then consuming carbon dioxide during photosynthesis. The rate of carbon dioxide growth was then a measure of our situation, a clear measurement of the rate at which our excess production of carbon dioxide was outrunning the biosphere's ability to either use or sequester it. Since oxygen and carbon dioxide are essentially paired opposites, this also meant that we were consuming oxygen more rapidly than plants could make it. I started calculating; I quickly concluded oxygen had to be dropping already, and that it could be measured. I called the Mauna Loa observatory, where the atmosphere is monitored, and asked if they had measured oxygen. It was June 1998 . . .

"Yes, we just finished measuring it with Scripps Oceanographic Institute a few weeks ago," said an official there.

"Well, how does it compare with the standard atmosphere numbers from 1958?" I asked.

"Oh! It's down!" he suddenly exclaimed

"How much? A few parts per million?"

"It's not a health risk! It won't affect anyone's health!"

"So, it's small?"

"Figure it out for yourself! We're burning it faster than we can make it!" He suddenly recovered his composure and told me to call Keeling at Scripps, then hung up.

Christy Booker moved away from St. Louis for two years during high school, and something miraculous occurred. During the entire time she was away, she was healthy in what has otherwise been a lifetime of serious and recurring problems with asthma. For the two years she lived in Philadelphia she never used her medications and she never had an asthma "attack." While we can't be absolutely certain what the difference was, Christy herself guesses that it was the air quality.

Many people with asthma are very sensitive to air pollutants, so, not surprisingly, when Christy was tested for allergies in early childhood, the tests indicated that she was allergic to a number of airborne pollutants. Ironically, Christy never needed to consult the various published and broadcast air quality reports. Why would she consult those sources when her body was already telling her exactly what the air quality was? The day's pollution ratings on television simply confirmed what she already knew. And since her body was such a sensitive instrument, highly efficient at detecting subtle changes in air quality, when she claims that her breathing was better when she left St. Louis because the air quality in Philadelphia was better for her, we'd have to consider that as a serious possibility. It was as if she had moved to another planet.

Currently, there are six major air pollutants that are monitored by the Environmental Protection Agency: carbon monoxide, ozone, nitrogen dioxide, sulfur dioxide, particulate matter, and lead.[13] Carbon dioxide, like hydrogen, nitrogen, helium and oxygen, is considered a background gas, and as such, is not currently considered a "pollutant." Most researchers consider carbon dioxide's only real potential for damage is in its role as an ever-increasing greenhouse gas, which can lead to more global warming. However, as this gas captures ever-greater percentages of our breathing space, isn't it reasonable to review its status as a "harmless" background gas? According to Wayne State University School of Medicine, while carbon dioxide is rarely encountered by itself in toxic concentrations, it may exacerbate the toxicity of other poisons, such as carbon monoxide and depleted oxygen.[14] While the effects of the very small atmospheric oxygen inventory depletion (OID) and of increased atmospheric carbon dioxide pose no known direct dangers to human health, these potentially hazardous trends cannot be overlooked.

There is little agreement about how much carbon dioxide is too much for human health. While the U.S. Occupational Safety and Health Administration sets the maximum safe limit on long-term exposure (time-weighted average) at 10,000 parts per million, which is a 1 percent concentration in the atmosphere, the American Conference of Governmental Industrial Hygienists set their safe maximum standards for long-term exposure at half that level.[15]

The importance and complexity of carbon dioxide in the human body has long been recognized in respiratory medicine. The classic work "Carbon Dioxide" published in the *Cyclopedia of Medicine* describes its importance this way: "Carbon dioxide is the chief immediate respiratory hormone."[16] More than the need to inhale oxygen, it is the biochemically-induced drive to expel carbon dioxide that stimulates our breathing mechanism.

Carbon dioxide at slightly elevated levels may also negatively alter the ability of our bodies to utilize oxygen within our cells. A study by the group of European scientists who made this discovery warns, "The results are probably significant for the analysis of important bio-ecological problems, such as the increase of carbon dioxide concentration in the atmosphere and its effect on humans and animals."[17] Even slightly increased levels of carbon dioxide change our normal breathing.[18] And not everyone reacts to increased levels of carbon dioxide in the same way. For example, in one study in which three men were asked to inhale a 3 percent concentration of carbon dioxide for three hours, two seemed to acclimate and one did not.[19] Will this be true at the tripled atmospheric concentrations levels (about 740 parts per million) now predicted for Earth?[20]

As you might imagine, there is a great deal of concern about the proper mix of gases aboard vehicles in space. For example, the near tragedy of Apollo 13 was due to increasing levels of carbon dioxide brought on by equipment failure. At this point, while studies have concluded that values around 1 percent carbon dioxide are acceptable for human habitation over several weeks in space, they point out that life science experiments on board may be sensitive to carbon dioxide at that level.[21] Imagine how it will be when every living thing on the planet—including those of us not up to the physical rigors of space travel—is subjected to higher levels of carbon dioxide. Could children and adults with asthma be "sensitive life science experiments" aboard our own spaceship, Earth? And what about all the other living things? Aren't we, in effect, conducting a dangerous atmospheric gas experiment right here, now?

Finally, a lungful of carbon dioxide in higher than normal levels of concentration can make us gasp and feel like we are choking. Anyone with asthma knows that gasps are not good things. Like the wrong food for a person with a food allergy, a gasp can easily trigger the body's self-protective asthma response.

If, at one level of concentration, carbon dioxide is a harmless background gas, and yet at another is a deadly toxin, at precisely what level of concentration does it become a problem? As long as the answer to this question remains unclear, should we simply accept the conventional wisdom that annually increasing levels of carbon dioxide, combined with falling levels of oxygen, pose no health problems for us? We are interested in the medical answers to this question,[22] not premature reassurances from atmospheric experts who aren't publicly forthcoming with important facts.

For most people, one of the problems with trying to conceive of a potential danger from carbon dioxide is it that is a challenge to think of an invisible gas as having physical weight. So, while the figures for greenhouse gases are routinely given in tons, it doesn't mean much in our ordinary sense of understanding. So let's convert the weight to volume in order to visualize just how much carbon dioxide that is.[23] Here goes:

A total of 7.1 petagrams (7.1 billion metric tons) of carbon dioxide was released into the atmosphere worldwide during 1997.[24] If it were all put in a 3-meter-high container (and the carbon dioxide was at 100 percent concentration), the container would be 700 miles square. You would have to drive for nearly 12 hours at 60 miles per hour before you reached the end of just one side. This volume would cover the entire U.S. eastern seaboard to a line 200 miles inland and a depth of 10 feet.

However, that doesn't even come close to giving you a sense of how potentially toxic this huge volume of invisible gas is. In order to understand that better it is helpful to know that the rules of the U.S. Occupational Safety and Health Administration (OSHA) state that carbon dioxide at a concentration of 3 percent (30,000 parts per million) is routinely considered very dangerous for humans to breathe for anything other than a few minutes.[25] At this level of concentration, the carbon dioxide we released into the atmosphere during the single year of 1997 would create an invisible toxic cloud that would cover a volume of 16.3 million square miles to a height of 10 feet. So, the 6,305 million tons released worldwide in 1997 would cover the Asian continent to a depth of 10 feet with extremely dangerous levels of carbon dioxide. (Keep in mind that the atmospheric carbon dioxide in this thought experiment would probably not have originated in Asia since the per capita emissions of carbon dioxide in China and India are a fraction of those produced in the US and Canada.)[26] Obviously, carbon dioxide mixes in the atmosphere to much lower levels of concentration and much of it is cycled into plant life and oceans. However, these examples will give you an idea of the tremendous amounts involved.

Soon after the release of the three MGS images of Cydonia, Dr. Horace Crater, Dr. Mitchell Swartz, Vince DiPietro and I were in Boston for the spring 1998 AGU meeting. Together with Mark Carlotto, Stanley McDaniel and Harry Moore, we'd submitted papers on the Cydonia region of Mars before the NASA decision to take images. In the scant few weeks since acquiring the new images, we'd worked feverishly to refine our understanding of them.

When JPL released the first new Cydonia image, it had its right and left borders reversed and had such poor contrast that it looked like a footprint in a snowbank. Its ineptitude would have produced a fail mark in any entry-level college image-processing class. One expert on the onboard camera system commented that it appeared that the "gain" had been set low to acquire the image. Whatever the cause, the results were almost completely unrecognizable and soon became known by our group as the "catbox"—

for reasons that don't need explaining. In the same afternoon, after releasing this astonishingly poor-quality image, certain non-NASA employees of JPL had announced to the national media that this proved the area was not artificial, and then announced that the whole investigation could be dismissed. Then, at 5:30 Pacific Standard Time, just after the West Coast media had put the story to bed and hours after the "catbox" image had been flashed and panned on television worldwide, JPL replaced the "catbox" image on the official government website with a more recognizable version. Nevertheless, the image and story were out: "the Mars face is a pile of rubble." They were JPL; who were we, despite our 20 years of expertise on the subject in question, to contradict them?

Michael Malin, a NASA subcontractor at Malin Space Science Systems in San Diego, California, had captured the infamous image, since he alone controlled the camera aboard the MGS and all the images it took. A long-time Cydonia scoffer, he had taken the image begrudgingly, finally capitulating under tremendous pressure from both NASA and the public. This was the same Michael Malin who so many years before had been our annoying neighbor at the Case for Mars conference in Boulder, Colorado. Over the years he had fulminated against the "face" investigation from his website. It was obvious that he approached the question of Cydonia with an extremely closed mind. But, adversary or not, Michael Malin owned all rights to the camera and to the first six months' analysis of all images it took. Like the announcer on the old *Outer Limits* television series, he could "control the horizontal and the vertical"—which, perhaps accidentally, literally occurred. The catbox image, according to one of our measurements, was skewed 17 degrees in the vertical plane and 20 in the horizontal plane, and had such poor contrast that it made the 600-foot-high mesa on which the "face" resided look like a sand dune.

I toyed at first with the idea of withdrawing our papers from the AGU, but as we began to examine the newly rectified image closely we found that nearly everything we had hoped for, and more, was there. It was very weathered, but nonetheless, the face appeared anatomically complete, with two eye indentations, nose, and mouth. Around its base ran an incredibly smooth and evenly rounded symmetric platform. It could easily be

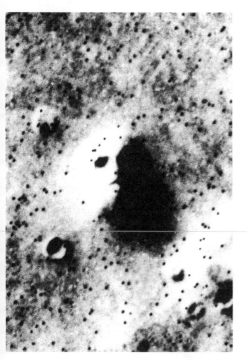

"Face" on Mars. Viking 1 Orbiter photograph of surface features in the northern latitude of the planet Mars. These formations resembling a human head measure 1 mile (1.5 kilometres) across. The photograph was taken on July 25, 1976. (NASA / Science Photo Library)

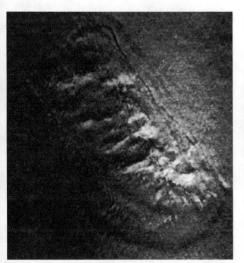

This is the Mars Global Surveyor image of the "face" that was broadcast on television and printed in newspapers around the world on April 6, 1998. It has become affectionately known as the "catbox" image. (NASA / JPL)

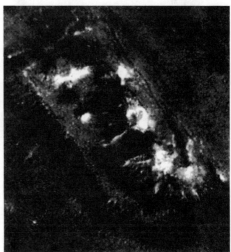

After processing to improve contrast and correct for smearing and skewing (the distortions created by camera movement and the highly acute angle from which the image was taken), this orthorectified image shows the same feature. (Dr. Mark Carlotto / NASA)

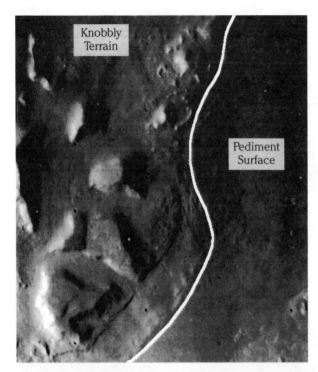

According to geologist James Erjavec and Dr. Brandenburg, the line of demarcation between the knobbly terrain and the much smoother pediment surface (shown here in the Cydonia region of Mars) may indicate the shoreline of an ancient ocean or other large body of water. (James Erjavec / NASA)

Egypt, Giza. View of the great Sphinx and pyramids. Could the face on Mars be a similar construction or is it simply an eroded mesa-like land formation? (Simon Harris / Robert Harding)

Joseph Priestley (1733–1804), the English chemist. He is best remembered for his discovery and examination of a number of new gases including hydrogen chloride, nitrous oxide, ammonia, nitrogen and carbon monoxide. He is often given credit for the isolation of oxygen although he never recognized its significance. (Science Photo Library)

The German bacteriologist Robert Koch (1843–1910). Koch is considered, together with Pasteur, as the founder of modern medical bacteriology. In 1905 he was awarded the Nobel Prize in medicine. (National Library of Medicine / Science Photo Library)

Svante Arrhenius (1859–1927), Nobel Prize-winning Swedish chemist at work in his laboratory. His achievements include his elucidation of the effect of temperature on reaction rates (the Arrhenius equation) and the discovery of the Greenhouse Effect. (Science Photo Library)

Florence Nightingale (1820–1910), medical reformer. She developed the use of statistics, charting, graphic displays and management procedures in the practice of medicine and demonstrated that using a rational approach in an arena that until then had been left unmanaged resulted in a significant reduction in the number of patient deaths. Her contribution to the evolution of the modern medical model will be useful in helping us to manage the complexities of our planetary environment today.
(Hulton Getty)

construed as a helmeted head, with what appeared to be symmetrical ornaments. A number of other intriguing new features appeared on the Cydonia plain as well. We resolved to proceed—to present our papers and offer our best analysis of an object that the world had already dismissed.

Few things are more terrifying to a scientist than the prospect of appearing ridiculous—to appear to be pressing a case after it has been disproved. However, we had always been a group of people who had swum against the tide. We were no longer not just Vincent DiPietro and John Brandenburg. The independent Mars investigation had grown considerably over the years. We were now The Society for Planetary SETI Research (SPSR), a highly credentialed group of scientists and engineers who had banded together to use the best available science to investigate Cydonia. (Our numbers included 14 Ph.D.s, several NASA subcontractors, a former member of NASA's Astronaut Corps, and the UN's Commissioner of the Conference on Peaceful Uses of Space.) Together we'd written a book and turned the tide of opinion; together we had contributed strongly to NASA's decision to order a U.S. space probe run by JPL to take pictures of Cydonia, something that JPL had adamantly refused to admit was important.[27] It was hardly surprising that when JPL released the images, they put their own hostile spin on them. This, however, was hardly an argument for abandoning our work.

In the aftermath of the image release and dismissal, the SPSR members rallied and encouraged one another and began the scientific work the tax-paying public thought they'd already received from government officials. We analyzed the images carefully—not just the face feature, but the whole region of Cydonia.[28] The sights we saw defined the term "unearthly." Fortunately, one member of SPSR was probably the best computer image-processing expert in the world: Dr. Mark Carlotto. Soon we had computer simulations of the Viking data rotated and illuminated under the same conditions as the MGS image of the face. It was a nearly perfect match. In addition, we had a rotation of the MGS image as a three-dimensional object, so that it would appear as viewed from above. Carlotto found that the object was 92–99 percent symmetric.[29]

In addition, our geologists began making geological discoveries. In one of the new images from Cydonia, geologist Harry Moore found what appeared to be water ice in the bottom of craters, an extraordinary and exciting finding, made even more so when another SPSR member noticed what appeared to be sunlight reflecting off the ice floors onto the far crater walls, as well as a similarly filled crater in another of the MGS images.[30] These two images were the first to show what may be bio-available water on the surface of Mars. Geochemist and SPSR member, James Erjavec, also found evidence of "wave-cut benching" suggesting that the face's landform may have been in the "midst of a paleo-lake" at some time in the Martian past, as well as what may be geologically recent evidence of precipitation-based erosion on the edges of the face mesa.[31]

So, it was a fine morning in Boston when we assembled at our appointed place in a great hall and put up our posters side by side as the AGU had graciously arranged. Immediately, a small crowd of bemused onlookers came to gawk at our presentation. Dr. Horace Crater and Dr. Mitchell Swartz tended the poster featuring the Mars images and geological findings. I held court in front of my poster about the CI meteors from Mars, and Vince DiPietro stood with the Cydonian Hypothesis poster.

I was very proud of our work. The whole group of presentations was impressive. Soon, onlookers stared in astonishment. Carlotto's work on the "face" was exhaustive and beautiful in appearance. There was a large reproduction of the MGS image, rotated to appear as if viewed from above, alongside the two Viking images. We had Harry Moore's work on ice in Mars craters, and there were a number of other Cydonian anomalies featured.[32] Between the two Cydonian papers, I presented the latest data on the CI meteorites. We now had an almost perfect match between the oxygen isotopes in the CI and those found in ALH84001, the "Martian life" rock. The paper Vince and I presented was of images of Cydonia, a review of geochemical data gathered by the MGS and Mars Pathfinder, and an analysis of the MGS image of the face, including the curious lack of contrast in the first version released.[33] Clearly, this presentation did not look like the last gasp of a defeated guerrilla band. Word apparently spread.

I stepped away for a moment to look at a poster done by a student, while Vince took a seat during a lull in the traffic. When I returned a solitary

figure was staring intently at our poster of the new MGS image. He was
fairly shaking with rage. It was the same scientist from the MGS mission
who had looked so troubled on CNN the night of the image's release. He
was staring at our rotated version of the MGS image. I had suggested below
it the name of "Cydone," the legendary founder of Cydonia in ancient Crete.
Stunned by his reaction, I tried anyway to greet him as cordially as possible.

"You're saying it's artificial!" he said angrily, pointing at the computer-
rotated image.

"No," I explained. "We only said it appears artificial based on our best
analysis. It does appear to have eyes, a mouth and a helmet. Look, it
appears to have two nostrils, too. Those were not visible in the Viking
images."

"No, you say it's artificial!" he asserted, becoming more angry. I became
a little irritated myself.

"No, the SPSR has always conceded that the face might be natural. Your
people, on the other hand, have never, not for an instant, ever admitted it
might be artificial," I replied. He only glared back at me in response. He was
obviously not a man used to being contradicted. Vince joined us in front of
the poster and chimed in.

"The grayscale on the MGS image appears narrower than we expected it
to be. That's why the contrast is so poor. Why is the grayscale on this image
narrower than on all the other Viking images?" Vince asked.

The man whirled on Vince, then he turned to me and said, almost
wearily, "We did the best job we could on the images . . ."

"Well," Vince continued, "with a grayscale this narrow, you will lose
effective resolution. Neighboring pixels will assume the same value and
look like one large pixel."

It was true. Vince and I had gone over this thoroughly before the
session. We called the phenomenon "rafting" of pixels, where in cases of
too narrow a grayscale range, neighboring pixels raft together at the same
gray value, thus detail is lost, as if the pixels were larger.

"You don't know what you're talking about!" the man screamed at
Vince.

Stunned by the sudden outburst, everyone at the neighboring posters turned and stared. The scene encompassed the whole Cydonian controversy in microcosm. Vince DiPietro, a talented technician and courageous explorer, but nevertheless, an electrical engineer without weighty credentials, vast institutional authority or lavish tax-supported salary, stood in face-to-face confrontation with the former chief scientist of JPL, in the hallowed halls of science. Vince looked at me in shock. I nodded fiercely.

Vince turned to the man and calmly, almost humbly said, "Yes, I do . . ."

"ARE YOU CALLING ME A LIAR?" yelled the man at the top of his lungs, raising his clenched fists. "I'LL DECK YOU, BY GOD!" he screamed. As everyone watched in horror, he lunged straight for Vince. I recovered from my shock and leaped in between them. Vince remained calm. To my great admiration, during this whole episode he said nothing in response, nor moved a muscle. The scientist bumped into me as I stepped in his way.

"Calm yourself, sir," I hurriedly urged and took his arm to lead him away from where Vince quietly stood his ground. Vince would later chide me for not letting the man hit him, but I still think this was better. Vince, in my eyes, became even more of a hero that day—immovable, not yielding to even the crudest sort of intimidation. In that moment the greatest impulses of humanity shone: a simple man standing in dignity for simple truth as he sees it, while a powerful gray-bearded sage screams and threatens. On that day, Vince DiPietro stood as a true man of science, as he had from the very beginning, over 20 years earlier.[34]

The incident—especially the reaction to our analysis—left me convinced that we had been looking at significant data all along. Our AGU presentation had thwarted a clear and conscious attempt to encourage the summary dismissal of the new MGS Cydonia data. We certainly weren't supposed to subject it to careful analysis.

Amazed by the turn of events, I said, "This means this thing isn't over, Vince. It means it's barely started."

The controversy in Cydonia still rages: after 20 years it cannot be settled with one image, nor three. More images need to be taken and will be. There will be a moment of truth for all of us someday. Meanwhile, some of us think we have viewed the glimmer of truth already. As people of science, it is our vocation to sit in the crow's nest of the great ship of human civilization and recognize distant things while they are still difficult to see; to give warnings, sometimes unheeded; and to point out great sights, even if everyone on deck is too busy too look. Sometimes we're right, sometimes we're wrong. Even if there once was a stone-age civilization on Mars— hopelessly trapped when the Lyot impactor killed its entire biosphere— perhaps humanity is not yet ready to find it. To recognize such a catastrophe would demand responsibility for all that is implied by such a discovery. But, if a tragic truth does lie on Mars, we are confident that humanity will one day stare it in the face and not flinch, even if our eyes fill with tears as we behold it, because, despite all of our flaws, our fears and our dreams, we yearn for, and ultimately can only live with, truth.

CHAPTER 11

Dying Earth

Mars died. The Lyot impact appears to have killed it. If so, then the Lyot impact transmogrified Mars—changing its former greenhouse-warmed state into its present one, a frozen husk. This is why Mars' liquid water erosion stopped after the Late Hesperian age. Water stopped flowing after the meteor at Lyot in the Early Amazonian age. Mars died, just like the dinosaurs. The exact time is difficult to ascertain because the Hesperian and Amazonian epochs are distinctions based on relative crater counts and we don't know the exact cratering rate at Mars, so absolute gauges aren't available. However, if it was four times the lunar rate, the Amazonian transition was 1.5 billion years ago.

The death of Mars by asteroid impact coincided roughly with a period of immense importance on Earth: the Precambrian explosion. It was during the Precambrian period that life suddenly became multi-celluar and began to evolve rapidly. Could advanced multi-celluar life have arrived from Mars aboard meteorites after Lyot and jump-started biology on Earth? Could extinguished Mars life have arisen phoenix-like on Earth? Could Arrhenius' support of the theory of panspermia be accurate? Herein lies a story.

It isn't easy for a modern theorist to be beholden to a scientist from a hundred years ago for our current findings, but it appears that Svante August Arrhenius' ideas about the biological processes of the cosmos, which seemed so entirely absurd for so long, may have more validity than we ever before considered. Could this same brilliant mind have been equally prophetic when alerting us, nearly a century ago, that by burning fossil fuels we were creating a greenhouse effect that was changing the thermal balance of the planet? Indeed, it seems now that he was—and his warnings were unheeded.

Sadly, one of the easiest predictions to make about global warming is that it will bring further war and turmoil to the Third World. The human race is the most unstable and sensitive part of the biosphere, particularly its economic and political systems. The Third World exists mostly in the tropics, and it is there that both unstable politics and the full force of global warming will combine. Environmental stress is one of the great destroyers of rational discourse, and one the most powerful triggers of desperate action. We are probably already seeing the effects of global warming on the governments of the tropics.

The first thing to go in the Third World will be peace, both domestic and foreign. Environmental stress will cause governments to abandon any pretense of democracy, or to disintegrate completely. One the most frightening things about the civil and national wars in Africa is their incomprehensibility to outsiders. But the problem of climate-induced anarchy is not limited to the Third World. Much of Miami, Florida was in a lawless state of looting and rioting for two days in August 1992 after the devastation of Hurricane Andrew. Only the National Guard was able to re-establish rule of law. As temperatures rise world wide, so will irrationality. Nation will rise against nation, not over anything as sophisticated as ideologies or oil, but over water and arable land. Turkey and Syria almost went to war in 1998; the pretext was terrorism, but the real cause was the water of the Euphrates.[1] Somalia was the most ethnically homogenous country in Africa, a continent known for its tribal conflicts, and was thus considered the most stable of African nation states. But by 1990, Somalia, which sits on the edge of a desert, was embroiled in civil war, there was drought, and soon the nation was dying. Somalia was too well armed to be pacified and too chaotic to be conquered. Even in a weakened state, Somalia still managed to bite off the fingers of the hands that tried to feed it. So the world, in a sense, abandoned Somalia. The people of Somalia are starving again but the news networks will not, or cannot, cover the wretchedness there. Hopeless misery makes poor programming. So Somalia starves in a back alley of the global village. Out of sight, out of mind. Is this the model for coverage of future global-warming-produced disasters? Will we just avoid looking?

Many chaotic places in the Third World lie on the edges of deserts or near the equator. As the world becomes hotter, the populations of these places, whose resources are almost down to zero, become more desperate and irrational. Disputes that formerly were resolved now often flare into bloodshed. In southern Algeria, where Muslim fanatics engage in pitiless slaughter, the southern regions, the fringes of the Sahara, have become so hot they are uninhabitable. In Rwanda, almost on the equator, where ethnic hatred led to genocide on a scale not seen since the Nazis, one of the problems was high population density. The fact that ground temperatures are rising naturally makes human tempers rise and probably makes overcrowding even more unbearable. In Rwanda, as has happened so many times before in our history, humanity's killer instinct was awakened as people savagely fought to reduce competition for vital resources.

One of the problems that makes any estimate of the real effects of greenhouse warming so difficult is that the global system is so complicated and so much of the greenhouse gas emission and absorption is mediated biologically. As has been discussed, an important and unpredictable part of the biosphere affecting climate is humanity itself. But the rest of the biosphere presents problems also. Because part of climate change is biological, it can display enormous sensitivities and unexpected couplings to other effects. This leads to nasty surprises—that global warming and ozone hole are coupled, for example. Ice crystals in the stratosphere are the sites of catalysis for ozone destruction. More thunderstorms in the polar regions owing to global warming increase the ice crystal supply in the stratosphere.

Another nasty surprise is the recent discovery that rainforest plants are not heat tolerant: if they get too hot, they die.[2] Who could have predicted this? It now appears some rainforests will become carbon dioxide sources rather than sinks. The rainforests are also being burned like garbage—like an annoying nuisance. A 1995 study published in *Nature* showed that, unlike our expectations that rainforests absorb carbon dioxide, in 1992, tropical regions between 30 degrees north and 30 degrees south were a net *source* of about 1.7 billion tons of atmospheric carbon. It was suggested that this "could reflect biomass burning."[3] This means the Amazon basin, most of tropical Africa, much of Southeast Asia and even some of southern Europe could become barren deserts if global temperatures reach a certain level—and Brazil and Southeast Asia continue their bold plans for economic development.[4] If this occurs, the present biomass of the Earth's tropics will rot or burn and become more carbon dioxide and a great desert will girdle the world.[5]

Already the Amazon desert is emerging in Rondonia. As far as the eye can see there is only red, cracked ground. The red dust fills the air and sky itself turns red. Ten years ago this area was jungle, then it was slashed and burned to form farmland. But the jungle soil is notoriously poor, so it was used for pasture. Once overgrazed it became desert. Now even the landless poor who created it have moved on. They have moved west, to begin the process deeper in the land that used to be the Amazon rainforest. We are recreating the Amazonian desert of Mars right here on the Earth. The only ludicrously absurd virtue in this is that, with the Amazon desert, the computer simulations of the greenhouse effect on weather will be simpler, and thus it will be easier to convince skeptics of the reality of global warming.

The surprises keep coming. As we go to press, ABC News reported on a new study, published in *Arctic and Alpine Research*, which shows that a temperature increase of only 2 °C causes significant releases of carbon dioxide from Arctic tundra soil, thus accelerating global warming.[6] Only 2 °C!

The Gaia principle is one of the most powerful and mystical ideas to appear in modern planetary biology. Essentially a principle of large complex ecosystems, it says that life alters its own environment to make it more hospitable to life, and that if outside forces alter the environment, life will react to minimize the negative impact of this alteration. A simple example: a rainforest spreads because its vegetation holds moisture, thus attracting more rain clouds which hover over it, circulating the moisture up from the ground, through the cycles of plant life, into the clouds and back down again as rain. According to the Gaia principle, the biosphere's response to increased carbon dioxide, and the global warming it produces, would be to increase plant growth so as to reflect more heat to the sky and absorb more carbon dioxide. Gaia also represents the hope that the biosphere will mitigate whatever negative changes humanity induces. However, on Mars, Gaia failed. Whatever life Mars had sustained could not overcome the forces that made Mars a frozen desert. The Gaia principle represents, at best, the tendency of a planetary biosphere to rebound from disaster. But its power is not limitless. Some form of the Gaia principle is probably at work on Earth now, but Gaia cannot take the place of intelligent foresight and action, nor overcome all of our excesses.

The Amazon provides one quarter of the Earth's oxygen. Rainforest is highly efficient in trapping light and thus promoting photosynthesis. Photons that penetrate the top canopy of the forest have usually two more canopies to go before they can be absorbed by the ground or reflected. Anywhere they go they run into something green. Grass lands, by contrast, allow most light to reflect or be absorbed by the ground; thus they are less efficient per acre in producing oxygen or trapping carbon dioxide. The fact that carbon dioxide and oxygen are exchanged one for one in photosynthesis and combustion-respiration was part of what led to the discovery that oxygen levels are dropping.

Oxygen Inventory Depletion (OID), the reduction in the Earth's planetary inventory or reservoir of oxygen, is probably the most alarming and potentially dangerous of all the global environmental problems we face.[7] The drop of approximately 50–70 parts per million, presently estimated since 1958, when it was last measured, is minuscule compared to the 210,000 parts per million that exist in the atmosphere. But oxygen is so vital to life that any drop must be taken seriously. OID must be stopped while it is still minuscule. The key to stopping OID is to phase out fossil fuels as rapidly as possible, to reforest great stretches of Earth, and protect the oceans. These seem like easy things to do but in fact they involve enormous economic challenges and are bitter medicine in an era that worships free-market economics.

Oxygen is one of the most reactive gases in nature and exists in the biosphere only through photosynthesis. Unlike carbon dioxide, it has no geochemical source, only sinks. It is an extraordinarily sensitive indicator of the health—or lack thereof—of the biosphere. Our oxygen supply is the most vital of the vital signs of life on Earth, and it is beginning to fail. The oxygen inventory is under pressure from two sources: first, the burning of fossil fuels is consuming it, and second, the land-based and oceanic plant life that produces oxygen is being destroyed. Any action that reduces the viability of the algae–plankton complex in the ocean reduces oxygen production, as does any desertification or deforestation.

Today, environmental platitudes, not religion, are the opiate of the people. They tell the people that a few degrees' rise in the global temperatures will mean resetting your air-conditioner to "high," that things will become more balmy, that change will be so gradual that we can all move our beach houses as the oceans rise. That is not the face of global warming that we are seeing even now. Global warming is a drought in the Sahel region of central Africa that never ends, forcing millions to move into refugee camps to live. It is killer hurricanes of force and destruction never before seen, destroying not just the economic life of swaths of Central America but of whole regions in the tropics. It is strange tropical diseases and parasites moving north to escape the heat, causing epidemics in populations that have never encountered them before.

Christy can tell you how vitally critical oxygen is. Since childhood, she has had to seek emergency room intervention for serious episodes of asthma, typically several times every year. Now, as an adult, she can tell us what the experience is like. Here's her description of an attack she had in December 1997.

> It can start any time of the day. This time it started early in the morning, at 5am. I had been up every few hours all night using my inhaler. I do that almost in my sleep without thinking of how long it has been. When it is late at night and I'm trying to sleep, I don't usually check my peak flow [testing the volume of exhaled breath using a peak flow meter][8] to see how bad it is. I don't need to test. I know I can't breathe. It feels like someone is sitting on me and trying to push all the air out. I can hear a wheezing sound as I struggle for breaths. I usually dream someone is choking me for a while before I finally wake up and realize I'm in trouble.

By the time the morning light comes, I try to remember how many times I got up and took my inhaler during the night. I'm only suppose to take it every four hours as needed. I take it again, because I can't remember when the last time was. I wear glasses and at night I can't see the clock. I find my inhalers by touch. I checked my peak flow and the reading was at 150. My "normal" readings are between 400 and 500. A reading of 150 is dangerously low.

I'm suppose to be in the emergency room when my readings are that low but I waited for 30 minutes to see what good the inhaler would do. When I rechecked my peak flow, I was only at 155; not a good response. Now I knew I'd have to do something else, soon. I woke my husband and we discussed what to do. I knew when I called my doctor he would put me in the hospital and I was terrified of the hospital. (Believe it or not, I am studying to be a nurse.) They usually have so much trouble putting an IV in my veins and they have to do that for some of the medications they give me. I dread the outcome.

By the time I called the doctor it was 6am and my peak-flow was 145 and dropping fast. I felt weak, dizzy at times, with too little energy to talk, dial, or think of what to say. I get very frustrated when that happens. This time, I dropped my glasses and broke them and I was in tears and this makes my breathing more labored and difficult which makes me more frustrated. It feels like you can't get any air in; you are taking breaths, but no oxygen is getting in for some reason. I am getting scared, scared of "what ifs." What if they can't get this under control? What if I don't make it to the hospital? What if I pass out? What if I die? That does come to mind every once in a while.

My husband took me to the hospital where we sat in the waiting room watching my peak flow slowly drop until they called my name and assessed my condition to see how severe the problem was. When they realized how bad it was, it had been a 20-minute wait and my peak flow was down to 135. I could hardly walk—didn't have the energy to do it. I just wanted to lie down. The Emergency Room staff gave me a breathing treatment and waited 30 minutes and then rechecked my peak flow: 175. It's still not good, so they did two more treatments 15 minutes apart and rechecked my peak flow half an hour after that: 250, an improvement at last.

The ER doctor decided to send me home and told me to come back if it got any worse. I felt better, but not great. They discharged me but my husband and I noticed my lips were blue when I left due to lack of oxygen. It took all my energy to get to the car. I was worn out. My peak flow was 200 as we headed off for home.

On the drive home, about a 20-minute trip, my peak flow dropped again to 150 and we returned to the hospital for a little vacation (haha). After treatment there I got a bit better, 200–250, but not very good. Later that night, I'd dropped to 100, which is definitely life threatening, and I panicked. I felt like I was literally going to die. I was so tired of trying to get a breath all day long, I just wanted to give up breathing. It hurt so bad, I didn't want to do it anymore. My back muscles ached beyond description from struggling so much.

They called my husband and began to talk about intubating me. [This process involves inserting a tube through the trachea.] I was terrified. The doctor increased my medications to an hour apart and my heart went crazy. It pounded so hard I could feel and hear it. Your heart really speeds up if you take a lot of medication, so I was put in the Intensive Care Unit for the next day.

After 12 hours of medications, I was finally, slowly, starting to improve. They reduced the medications to every 2–3 hours and I was moved to the Heart Unit for monitoring. When I finally got so I could breathe with the medications being administered only every four hours, I was moved again and became a general patient.

After five days in the hospital, it was Christmas Eve and I wanted to be at home with my husband and children. The doctor sent me home, unwillingly, but slowly I got better and didn't have to go back. I was out of work for two weeks and my breathing was about 250–300 for the first month after that. I took very high levels of medications for a couple of weeks and I then tapered off very slowly. It took a full month before I was back to my "normal" medications. It took two months to get my peak flows up to 300 every day. By three months, I was up to 400.

I breathe at about a 500-level peak flow now, but my smaller airways are not very good, even on a good day. I have some permanent damage from that attack. I still can't work out. I can't walk a long distance without needing my inhaler and having tightness in my chest. And if I get a cold, any cold, I am back in the hospital within a couple of days. I'm glad to be alive—my family needs me—but I may never be the same.

Christy Booker survived this life-threatening asthma attack. Many others are not so lucky and the numbers of people dying from asthma have been increasing dramatically. Among 5–24-year-olds, the asthma death rate nearly doubled between 1980 and 1993.[9] The death rate for black Americans is almost three times that of whites, and higher still for those black Americans living in cities.[10] But this is not just an American inner city phenomenon. The incidence and mortality of asthma is increasing worldwide. Even with the best medical care available, the scenario Christy has generously shared with us is played out hundreds of times a day—tragically, too often to a fatal conclusion.

For how to conduct a peak flow meter test, see Appendix B.

A small team of humans laboriously put on their protective suits and goggles. The suits were insulated and the goggles shielded their eyes and trapped moisture, so they could see. Once ready, they left the shelter of their vehicles and walked across the searing hot sand to the distant craters. Around them the sand, in surreal dunes, stretched everywhere to the far horizon, as beautiful as it was desolate, beneath an eternally sapphire sky.

The craters were large and poked out the sand dunes. The meteors that centuries before had created them had struck with such violence that the sand had fused to glass—as if at a nuclear weapon test site. The air waved and shimmered the scenery in the heat, its temperature a blazing 52 °C. As much as they wanted to study the site—after all, they had come so far to visit this forsaken place—the humans could not look for long. The suits did not work well and before long they hurried back to their air-conditioned vehicles. One of the party, overcome by heat exhaustion, began babbling incoherently in a mixture of English and Arabic.

The place where this occurred was not on some distant planet but on our own. This is a description of the ArRab Al Kahli, the empty quarter of Saudi Arabia. There is only the most hardy of vegetation there, and no water. The air temperatures will kill even camels. It used to be bad there, but conditions are much worse now. This is one of many zones on this planet that are uninhabitable because of heat. And they are spreading.

We are Mars-a-forming the Earth. While we were busy exploring our nearest neighbor, looking for life and finding disaster, we were tolerating the Marsification of our own planet. Evidently we still aren't getting the picture. Yet, it is so simple, really. Look to the night sky and find the red planet. Mars is Exhibit A. Observe.

> Mars once had an ocean, larger than the Pacific.
> Mars once had an atmosphere.
> Mars once had life.
> They are all gone.

B ad luck runs in streams through great disasters, and it floods through a catastrophe. The *Titanic* was traveling three weeks later than originally planned because its sister ship, the *Olympic*, had suffered a collision and needed its dry dock. The winter was very mild that year and when *Titanic* made its maiden crossing of the Atlantic, the Greenland icepack had begun to thaw and shed thousands of icebergs. The watchmen had no binoculars and it was a moonless night. The water was still, so no whitecaps girdled the iceberg to reveal it. The iceberg in question had recently capsized and thus exposed its darkest portion above water, and the ship was traveling much too fast. At dawn the *Californian*'s crew awoke and noted that much activity lay in the distance. They turned on their radio and discovered the largest ship in the world had hit an iceberg and sunk in the night, killing 1,500 people just 20 miles from where they sat minding their own business.

The mystery of the distress flares during the night was now solved. Captain Lord ordered the ship underway and cautiously moved to the site of the disaster, taking two hours. But a large passenger ship, the *Carpathia*, had arrived four hours earlier and had already picked up all the survivors. They pulled up beside it. The captain of the *Californian* offered to help carry some of the survivors back to port, but the master of the *Carpathia* refused, so the *Californian* had to content itself with picking up the remains.

One can imagine a scenario for global catastrophe that runs similarly. If the human race adopted a mentality like the crew aboard the ship *Californian*—as some urge, saying that both ozone hole and global warming will disappear if statistics are properly examined, and we need do nothing about either—the following scenario could occur.

The world goes on its merry way and fossil fuels continue to power it. Rather than making painful or politically difficult choices, such as investing in fusion research or enacting a rigorous plan of conserving, the industrial world chooses to muddle through the temperature climb. Let's imagine that America and Europe are too worried about economic dislocation to change course.

The ozone hole expands, driven by a monstrous synergy with global warming that puts more catalytic ice crystals into the stratosphere, but this affects the far north and south and not the major nations' heartlands. The seas rise, the tropics roast but the media networks no longer cover it. The Amazon rainforest becomes the Amazon desert. Oxygen levels fall, but profits rise for those who can provide it in bottles.

An equatorial high pressure zone forms, forcing drought in central Africa and Brazil, the Nile dries up and the monsoons fail. Then inevitably, at some unlucky point in time, a major unexpected event occurs—a major volcanic eruption, a sudden and dramatic shift in ocean circulation or a large asteroid impact (those who think freakish accidents do not occur have paid little attention to life or Mars), or a nuclear war that starts between Pakistan and

India and escalates to involve China and Russia . . . Suddenly the gradual climb in global temperatures goes on a mad excursion as the oceans warm and release large amounts of dissolved carbon dioxide from their lower depths into the atmosphere. Oxygen levels go down precipitously as oxygen replaces lost oceanic carbon dioxide. Asthma cases double and then double again. Now a third of the world fears breathing. As the oceans dump carbon dioxide, the greenhouse effect increases, which further warms the oceans, causing them to dump even more carbon. Because of the heat, plants die and burn in enormous fires which release more carbon dioxide, and the oceans evaporate, adding more water vapor to the greenhouse. Soon, we are in what is termed a runaway greenhouse effect, as happened to Venus eons ago. The last two surviving scientists inevitably argue, one telling the other, "See! I told you the missing sink was in the ocean!"

Earth, as we know it, dies. After this Venusian excursion in temperatures, the oxygen disappears into the soil, the oceans evaporate and are lost and the dead Earth loses its ozone layer completely.

Earth is too far from the Sun for it to be the second Venus for long. Its atmosphere is slowly lost—as is its water—because of ultraviolet bombardment breaking up all the molecules apart from carbon dioxide. As the atmosphere becomes thin, the Earth becomes colder. For a short while temperatures are nearly normal, but the ultraviolet sears any life that tries to make a comeback. The carbon dioxide thins out to form a thin veneer with a few wispy clouds and dust devils. Earth becomes the second Mars—red, desolate, with perhaps a few hardy microbes surviving.

In what was once Egypt, near a large but dried-out river bed, a group of pyramids and an eroded Sphinx confront the dead sky. In a distant future a passing probe from another civilization takes a picture, but most of the scientists who see it are skeptical that it could represent anything artificial and ridicule those who think otherwise.

The social, political and economic upheavals that are brought on by global warming are not something that will happen in the distant future. They have already begun. Indonesia is the present exemplar for global-warming-induced environmental disasters, because its collapse had repercussions that affected the world, but it has competitors: China, North Korea, Nicaragua, Honduras, Sudan, Siberia, Brazil, Haiti—with Florida and Germany chasing closely behind. For those who need a graphic illustration of the principle of environment as economy, Indonesia is a perfect case study.

In addition to being avid colonists, the members of the Dutch government in Indonesia maintained very good records of the temperatures back in the 1860s. These records were unearthed by UNESCO in 1992 and used to establish a scientific basis for what Roger Harger, coordinator for environmental programs at the Regional Office for Science and Technology in Southeast Asia, found was evidence of a 1.66 °C increase in the mean monthly temperature of Indonesia by 1990. This is far greater than the 0.58 °C increase the World Meteorological Organization announced that Earth experienced between 1961 and 1990, and as such, it may be very significant.[11] While this may not sound like a huge increase, it was all that was needed to set off a tragedy that continues to unfold to this day. Already, back in 1992, Indonesia, a tropical country with 353.4 million acres of tropical rainforests, was suffering drought. The average number of dry months increased from 4.4 per year to 5.4, with a dry month defined as one with less than 4 inches of rain—that being the amount needed to sustain rice production on the islands.[12]

By 1994, scientific research was making a connection between the droughts in Indonesia and overall changes in the Earth's climatic patterns. A study noticed that there had been an increase in intensity in the El Niño–linked dry spells, particularly in the Kalimantan region where there were associated economic losses, particularly in small rural villages.[13]

Indonesia had been feeling another kind of heat as well. During the elections held in March 1997, there was a growing opposition movement despite the threat of armed intervention by the military. Riots marred the parliamentary election of President Suharto, in what he called "a feast of democracy," as he began his sixth five-year term.[14]

In the summer of 1997, Indonesia went up in flames after years of climatically-induced long-term drought. By the end of July, fire had destroyed 1.06 billion cubic feet of timber and 5,724 acres of forest. Many of the fires (90 percent according to the Indonesian Environmental Minister) were started by large forestry and plantation companies and the government-sponsored transmigration (rural resettlement) program. The remaining 10 percent were caused by traditional slash and burn harvesting and land-clearing methods. A thick "haze," as the smoke was euphemistically called, began to issue from the forest fires on Sumatra and Kalimantan, hanging in the air over both islands and that of their neighbors, Malaysia and Singapore.[15]

By the end of the first week in August, it was estimated that there were more than 600 forest fires in Indonesia. Smoke was so dense on the islands of Borneo, Java, and Sumatra that it impeded air traffic and caused the emergency landing of a Boeing 737.[16]

While only a little over 5,700 acres had been involved at the end of July, by mid-September over 740,000 acres were burning out of control in what had become the worst Indonesian drought in over 50 years. Twenty million Indonesians were choking as increasing amounts of smoke created a serious health hazard. Government leaders began to question for the first time why logging companies were allowed to burn off unused wood from forestry operations that could instead be recycled. Clear-cutting had damaged watershed areas, leaving barren grasslands vulnerable to erosion and silting problems. As a testimony to widespread crony capitalism and corporate abuse, overshadowed by a long-standing authoritarian regime, of 120 million forested acres that had been designated as legally protected, only 49 million remained.[17] Many of the hundreds of environmental abusers who used fire as a quick method for clearing land were companies, including rubber and palm oil planters, foresters, and extractors of gold and coal. In a bitter paradox, those who sought to expand the economy by destroying the environment inevitably destroyed the economy as well, as is, ultimately, always the case. Ironically, many of the companies who devastated Indonesia for profit were receiving international funding for their efforts from purportedly "environmentally enlightened" countries such as Canada, New Zealand, and Finland.[18, 19, 20]

The drought continued through September, and Indonesia attempted to make rain by seeding clouds, but without success. Malaysia sent 1,210 firefighters to help extinguish the flames. The increasing smoke became a problem throughout Southeast Asia, prompting authorities in the Philippines to worry about the smoke triggering asthmatic reactions in that country. More than 32,000 people on Sumatra and Borneo suffered respiratory problems. Eye infections became common. With air quality registering well above the "hazardous" level, a state of emergency was declared in Borneo, where U.S. Embassy personnel, feeling the ill effects of the smoke, abandoned their posts, leaving only a skeleton staff in rotation. Citizens were asked to wear protective masks, yet these were described as useless, since they held back only 10 percent of pollutants. On September 23 in Kuching, the capital of the state of Sarawak, the government's Air Pollutant Index, which measures levels of sulfur dioxide, carbon dioxide, nitrous oxide, lead and dust particles, hit a high of 839 on a scale which indicates that a reading between 201 and 300 is "very unhealthy" and between 301 and 500 is "hazardous." The Mount Rinjani National Park on the island of Lombok, overlooking Bali, burned, the choking smog sending tourists home. Worse yet, a jet carrying 234 people crashed into a smoke-enshrouded mountain, killing all on board. Literally tens of thousands of people sought treatment for skin and respiratory problems.[21] The government considered evacuating millions of people, but since no one knew when the monsoon rains would finally come to end the fires, the evacuation of so many people seemed futile.[22] Of course, people weren't the only ones affected. The fires destroyed the habitats of a number of endangered species, including orangutans and spotted leopards—who undoubtedly suffered from the effects of smoke inhalation as well.[23]

Hell in a handbasket arrived in 1997, and Indonesia was going there quickly. If investors had been watching the natural resources of Indonesia burn unabated and realized that this was the real wealth behind their investments, they might have smelled the smoke earlier and looked for the exits. However, investors were watching the numbers instead—which still looked good at the tail end of October, 1997—just before the entire Asian economy collapsed in a lethal, economic domino effect. Indonesia, although perhaps not the absolute first to fall (Thailand's currency was the first to go), was certainly the largest, fastest and deadliest domino to crash and burn, sending the world's economy reeling.

Until then, the economy of Indonesia had looked as healthy as the rainforests did before the drought. Wages were rising, inflation was low, exports were projected to rise 14 percent. "There were no obvious warning signals of the kind of catastrophe that was about to hit Indonesia—at least [none] that we were watching," said Dennis de Tray of the World Bank office in Jakarta. Of course, a corrupt corporate and political system operating in an unregulated economy also regularly fueled the flames of the economic disaster by releasing an ongoing stream of deliberately overly optimistic financial statements.[24]

So, even as the government stood by and ineffectively watched Indonesia's fires rage out of control, and as its neighbors complained bitterly about the smoke they were forced to breathe, inevitably it was the Indonesian economy which began to burn as well. For the first time in years, growth began to fall and the specter of increased unemployment grew in a country chronically plagued by underemployment. The value of the currency plummeted 30 percent and suddenly, the 50 billion dollar debt the Indonesian government owed foreign investors, never a concern before, looked unmanageable. Even the President for life was starting to look like a liability.[25] Of course, global warming is visited upon our real world with all of its complexities and liabilities, not on some computer model world where political and economic realities don't enter the picture.

By the end of October, the stresses of environmental disaster were beginning to show up in Indonesia in ways other than currency collapse, drought and fires. Crops failed and about 500 people died as a result of food shortages and disease in what had not long before been a land of plenty. Devastatingly poor rice harvests were as much as 40 percent below normal, and corn, peanut, fruit and vegetable harvests were badly affected as well.[26] With the country still in flames, political instability growing, a repressive military standing at alert, and prices rising because of the drop in currency values and increased unemployment, it was clear that adding food shortages to the volatile mix could easily lead to an environmental/economic/political meltdown.

Seasonal rains finally arrived in November and most of the fires were extinguished. However, along with the monsoon rains came flooding, causing hundreds to be evacuated from their homes.[27] In addition to the massive economic losses due to drought and burned acres of lumber, it was estimated that costs of over $1.3 billion in health and other environmental factors were caused by the pollution, and that neighboring Malaysia and Singapore lost at least $360 million because of lost industrial production and the drop in tourism.[28] The full cost will never be known because the full long-term damage to human health and the environment cannot be fully assessed.[29]

· However, the reprieve from fires didn't last long and by the end of February 1998 the drought had resumed full force and Indonesia was in flames again. In Borneo, 34,600 acres of forest were burned, and another 9,900 acres in Kalimantan. Now cash-strapped, Indonesia sought international aid to try and head off the worsening problem. Monitored by satellite, hundreds of fires were counted; some were rekindled from smoldering peat or underground seams of coal, ordinarily too wet to burn, that had never really stopped burning in 1997 because of the drought.[30]

Soldiers wielding hoes and shovels attempted to dig firebreaks. Villagers tried to douse the flames carrying buckets of water from creeks and rivers, which were already at an all-time low due to months of drought. The haze of smoke was increasing and the possibility of another summer of fire loomed. By March, up to 1,000 fires—many still deliberately lit by farmers, and plantation and timber firms—were ablaze. The toxic mix of greed, ignorance and global warming sent sparks from tree to tree across the jungle canopy, starting fresh blazes. Just like the U.S. farmers in the Dust Bowl of the 1930s, who continued to plow the dust and replant over and over despite the clear failure of their methods of agriculture, denial can make all of us behave irrationally. Again, trees weren't the only living things in the rainforest to die. Drought and fire so reduced the flow of the Mahakam river that it endangered over 200 rare fresh-water dolphins. At the same time, reduced forests continued to cause orangutan populations to decline precipitously.[31]

The human population of Indonesia was beginning to feel endangered too. Inflation began to soar at rates of 30 percent or more. Milk was prohibitively expensive. Workers in an auto-body repair shop went on strike demanding that their 80 cents per day wages, inadequate to meet even the most minimal living costs, be raised to $1.50 to keep up with the inflationary prices. Management responded by folding the shop and firing the 40 workers. The majority of Indonesia's 200 commercial banks were insolvent and credit was unobtainable.[32] Chinese shopkeepers were now the targets of violent protests. President Suharto rejected any proposed reforms, even when the request arrived directly from the U.S. presidential representative, former Ambassador to Japan Walter Mondale, and the International Monetary Fund.[33] By May, remarkably, the U.S. backed a billion-dollar IMF bailout, despite no formal agreement to reforms and evidence that Suharto's cronies were resisting any actions that threatened to break up the monopolies they controlled.[34]

The fires burned on. Brunei was blanketed with smoke from Indonesian and Malaysian fires. Flights were canceled and the airports shut down periodically when visibility was limited.[35] The fires spread, reaching further and further into the virgin rainforest. Over 700,000 acres burned between January and the end of March. An alarmed United Nations convened an emergency meeting in Geneva to plan an intervention in the fire, hoping that the U.S. and other nations would pay $10 million to train and equip 1,000 fire fighters.[36] By May, Indonesia, Thailand, Malaysia, and Vietnam all were suffering heat waves, drought, and fires. In the Philippines, 200,000 families left remote mountain villages and descended on the cities asking for food. Food shortages throughout Southeast Asia were common. Rare plant species died, animals burned or starved. Environmental agencies estimate that the process of logging and conversion to crop plantations left only 1.5 million square miles of rainforest out of Asia's original 6 million square miles. And much of what little was left was now vanishing in smoke.[37]

In May the Indonesian poor rioted when fuel prices jumped 70 percent overnight, the result of removing subsidies, under IMF pressure. Looting and arson were widespread. Between the 13th and the 15th of May, over 500 people died in an orgy of violence. On May 19th, upward of 30,000 student demonstrators demanded that Suharto step down. Even the middle class, who days earlier would not have dreamed of demonstrating in front of the military, took to the streets.[38] Protests lead to the resignation of the President of 32 years on the 21st of May. His protégé and vice-president, Bacharuddin Jusuf Habibie, was elevated to the presidency. The Chinese minority, some of whom were shopkeepers and viewed as "too rich" by the Muslim majority, were the major target of the violence. Those Chinese who were too poor to fly out of the country armed themselves with whatever they could get their hands on and attempted to barricade the streets. Over the ensuing months, it was discovered that the rape of Chinese women by the Indonesian military had been widespread during the riots.[39] However, along with their poor and sometimes hungry oppressors, even the Chinese shared a mutual hatred of Suharto, and, like the majority of Indonesians, blamed him for the economic woes.[40] During the rioting, the U.S. Federal Reserve Chairman, Alan Greenspan, warned that the Asian financial crisis remained highly volatile and that the impact of the meltdown in the currency and stock markets was just beginning to be seen.[41]

The Indonesian economy died. No one was buying anything but rice and oil. No one was selling. Factories were closed for lack of raw materials. Fuel costs made distribution impossible. Prices soared and the remaining rice supply plummeted. Private companies owed $78 billion in debt and 90 percent of the companies traded on the Jakarta Stock Exchange were insolvent. Inflation was at 50 percent and expected to go up. The currency continued to fall. Unemployment hovered as high as 70 percent.[42] All of this was on top of at least $4.4 billion worth of damage caused by the fires.[43]

Alan Greenspan had been correct when he warned that political and economic trials for Indonesia might be just beginning. Mob violence continued, now directed against both Christian and Chinese minorities, as a once pluralistic society fractured along ethnic and religious lines. There was continued looting by hungry Indonesians, many of whom had been productive citizens a year before. Tens of millions of Indonesians sank into poverty. La Niña, the follow-up to an El Niño cycle, threatened flooding again during the monsoon season. Corruption also continued and the country periodically exploded in riots, even as political reforms were introduced. Murder had become commonplace as anarchy reined.

It's shocking what a 1.66 °C mean temperature increase can do, isn't it? Global warming does not occur in a scientific laboratory nor on a computer screen witnessed only by those eager to perfect their version of the global atmospheric model. The textbook on global warming may be written in universities, but its history is being burned in human lives in Indonesia and places like it. Indonesia is a textbook case of how to mismanage a drought, how to mismanage land use, how to mismanage a "democracy." The result of that mismanagement will obviously contribute to the problem of global warming for all of us, as the fires pour greenhouse gases into the atmosphere. Human factors are a big piece in the equation—but then they always are with global warming. Global warming, as we are experiencing it now on Earth, is *caused* by humans, through ignorance, greed, entitlement, over-consumption, over-population, unconsciousness, denial, failed policies, cronyism, selfishness, and shortsightedness, or by simply perpetuating the earth-damaging traditions into which we were born. All of these can be deadly, but perhaps deadliest of all is our denial of the part each one of us, personally, is playing—thinking all the while that it is someone else who is creating the problem. While we may not be personally slashing and burning or having too many children, we are, predictably, doing something that is adding to the carbon load of the atmosphere and thinking that we, unlike those others, are justified in doing so. We are all engaged in this process and we must find the way out together—because we're all *in* this together. The reality is this: if the ship you're on is sinking, it doesn't matter that it isn't your fault.

Indonesia did not burn in isolation. It was just the worst of the multitude of places which burned during the summer of 1998. At various times a ring of fire circled the globe as rainforests burned: 2,150 square miles in Central America; 1,500 square miles in Mexico in the worst drought in 70 years; and 20,000 square miles in Brazil, an area about the size of Slovakia. Fires also engulfed huge tracts of Siberian Russia, Canada, Kenya, Rwanda, Tanzania, Senegal, the Congo, and Florida in the U.S. Many of the world's scientists conceded that there was something more than the mere seasonal climatic variations of El Niño at play, and they began to suspect, as *Time* magazine said, "that something more fundamental—and frightening—is happening."[44] That something is the global warming produced by the excessive greenhouse gases from humanity's continual burning of oil, coal, and vegetation for decade after decade. Tragically, in a fiery feedback loop, all the burning was adding even more greenhouse gases to the atmosphere, making a future with even greater problems even more of a certainty.

And those were just the fires. We will spare you a full recounting of the record-setting heat, flooding, hurricanes, tornadoes, and droughts that befell our Earth during 1998, killing over 50,000 people and inflicting over $89 billion worth of damage—more than occurred during the entire decade of the 1980s (even adjusted for inflation). Also, in 1998, almost unbelievably, over 5 percent of Earth's entire human population became environmental refugees, as over 300 million people were forced from their homes by climatic disasters.[45]

The millennium ends on an ominous note: in what researchers are calling a "startling revelation," average temperatures for 1998 were the hottest in the last thousand years.[46] The 1990s were warmer than any other decade in the past millennium.[47]

If Mars is Exhibit A, what is Earth?

The Earth has oceans as wide as the Pacific.
The Earth has an atmosphere with oxygen to breathe.
She is the physical abode for every living thing we have ever loved.
She is ours, to love and to cherish for as long as we both shall live.
She's the gift we leave our children,
and the only home we have.

Earth is dying.
We must save her.

CHAPTER 12

The Enterprise

My name is Isabella Estancia Perez,
representing Ecuador, and my nation's
most important export is oxygen . . .

A voice from the future

Early in the morning of April 15, 1912 the *Carpathia*, a 30,000-ton passenger liner, was surging slowly through the waters of the North Atlantic at a steady 10 knots in a southeasterly direction from New York to Gibraltar. It was just past midnight when the CQD (Come Quick Danger) message came through on the radio. A ship had hit an iceberg at North 41° 46', West 50° 14'—60 miles northwest of the *Carpathia*. The news that the ship in trouble was the *Titanic* was met was astonished disbelief by the crew—she was after all the largest ship afloat at the time. But once the *Carpathia*'s captain was informed, he immediately organized an emergency operation and ordered that a radio message be sent to the *Titanic* telling her help was on the way at all possible speed. From his 20 years' experience at sea, Captain Rostron knew that lives were at stake and there was no time to lose.

Then began a superhuman effort on the part of the "Black Gang"—the ship's stokers—as they made "a dash" to save the men, women and children aboard the supposedly unsinkable *Titanic*. Imagine the scene: The stokers who had been sleeping were aroused to come to the aid of the shift on duty, and in a coordinated effort, lines of men shoveled coal and slung it deep into the white-hot mouths of the furnaces, urged on by the chief engineering officer's shouts of, "More speed. The Captain wants more speed!"

Elsewhere on board, preparations were made for the rescue. Nets and lights were hung off the side, rope ladders attached to each gangway, and the launch boats made ready to pick up survivors. Watches were doubled to look out for icebergs. "We can do it! *Carpe diem, Carpathia!*" shouted the captain, as his crew fought to seize the day from the frozen Arctic night.

Soon the *Carpathia* was turned around and sailing at full speed towards the stricken *Titanic*. With her boilers near to bursting point, she raced through the dark sea, avoiding icebergs of the kind that had just sunk the largest, most modern ship ever to sail. Yet the *Carpathia* bore on relentlessly, never slackening speed. There were lives to be saved . . .

The 20th century, and the second millennium, are drawing to a close. A new century, a new millennium and, hopefully, a new age dawns. In our journey across time and space we have searched for truths that would help us at this threshold. To live is to be exposed to danger; to experience apprehension is a condition of existence. To follow reason and awareness, we must be responsible. We have seen how the storms of Mars first warned us of our collective madness. You now know how Venus became an inferno and how the Shoemaker-Levy impacts caused atmospheric firestorms as big as the Earth to erupt on Jupiter. We have been warned of potential threats to come. We cannot deny what we have seen, and we are responsible for this knowledge now. The planetary environment we possess is precious, it is fragile—and events within our control, both human and natural, can destroy it utterly. We now know that a whole planet can die, and that we can expect no agency to help us if we do not help ourselves. With this knowledge, we become responsible. We are responsible because we can act.

Carbon dioxide, that most rugged and simple of molecules, is the substance that remains when the inferno has burned itself out, when all that can draw breath breathes no more. Carbon dioxide is the lowest common denominator, the substance that remains after all the fine things, chlorophyll, water, and hope, have been lost. It inherits planets after all planetary disasters have run their course.

We are in the beginnings of such planetary disaster now. Like passengers

244

on the *Titanic*, we've only just felt a slight bump and noticed that the ship has stopped. Here in the First World, we note that the ship's officers seem unusually active—maybe a bit too cheerful and reassuring—and they seem to be locking all the doors between steerage and the open decks. This is how a disaster looks when it begins. We see the evolving probability field of the catastrophe around us.

The opening of the ozone hole, so unexpected, so massive and so full of portent, was the occasion for a worried meeting by the leadership of humanity in Montreal, which concluded on September 16, 1987. This meeting resulted in a treaty that we obey even though the US national legislators, paying homage to their donors, failed to ratified it. The effect of the Montreal Accord has been to raise the price of air conditioning and create several new industries, one of which is professional skepticism. The meeting in Montreal also saved our lives.

An attempted response to the disaster of global warming has now begun. The world's leaders met in Kyoto, the beautiful ancestral capital of Japanese culture and a city so precious that, even during the horrors and hatred of the Second World War, it was never bombed. Having survived the fire storms of war unscathed, Kyoto stands as testimony to our hope that things will eventually get better, that beauty is not lost and that peace will prevail. At Kyoto, by December 10, 1997, the world community had drafted a treaty to limit emissions of greenhouse gases. Adopted by over 160 nations on December 11, 1997, the Kyoto Protocol set binding limits on emissions for the developed countries that are most responsible for current levels of greenhouse pollution and it created significant incentives for developing countries to control their emissions as their economies grow. Hashing this agreement out was a rancorous process, full of all of the irony and tensions that shape culture and history—humanity at its worst and best. Now the response to the response has begun and it is far more severe than the one that greeted the Montreal treaty regarding ozone.

The pose of scientific detachment is wearing thin. Whole industries, the lifeblood of whole nations, more wealth than can be imagined, and our whole way of high-powered living has been called into question. We are no longer up in the ozone. When we discuss global warming we are down here

245

on the ground where we live. Far more than the cost of air conditioning will have to change before we are done, if we are to solve these problems and to live.

There will be a chorus of voices, some speaking from the depths of conviction, and others, well-paid ones, who will adamantly insist that the carbon dioxide level is not affecting our climate. Yet others will be equally adamant that carbon dioxide will affect our climate but that the effects will be benign or even helpful—things will just get a bit balmy, plants will grow better, a bit more rain will fall. They will say these things even as the Nile and the monsoons fail and the Sahara encoaches like a ravenous monster and swallows the Sahel. Some nations—those who rely on fossil fuel exports or imports for their economic life—will go further and forbid any mention of greenhouse warming and its connection with fossil fuels. To even suggest there is a problem may be life-threatening in some parts of the world. Do you know that it can be a death sentence to speak too loudly about the damaged rainforest in parts of Brazil? However, our *Titanic*, our "unsinkable" Earth, has begun to list noticeably. The signs are everywhere that things are out of balance. The hurricanes are stronger, the winters milder, the snows deeper, the summers more ferocious, and the zones of human habitation are shrinking as the deserts spread.

A terrible synergism of disaster is already at work. The complex system called "climate" is running amok because of increasing carbon dioxide, while at the same time, oxygen, the "other gas" involved in the combustion of fossil fuels, is losing concentration levels in our atmosphere. We are talking *oxygen*, the gas that we breathe in to fire every cell of our bodies—not the carbon dioxide that we breathe out as waste, but the stuff we need to sustain the process called "life." The decline of oxygen is tiny, but easily measurable. Its decline may have been noted years ago, but its significance was immediately minimized. In a bow to its emotional implications, the data was suppressed—or, given the human ability to distance or deny—maybe even repressed.

The decline in oxygen concentration means the beginning of the end for fossil fuels. To continue to burn them at the present rate, to contemplate that we will industrialize the Third World based on fossil fuel use, to consider that the world's rainforests are just idle land to be burned and farmed, is to participate in an act of environmental genocide and self-immolation. Some will insist that even though the world's supply of oxygen is going down, the amount is too small to be important. That is nonsense. It is important. On the course we are on, it will continue to fall. Finally, it will plummet like a stone. The decline in oxygen is important because it shows where we are going. It is akin to the canary falling off its perch in the coal mine, or the frantic call from the crow's nest that an iceberg is dead ahead.

When we hear the flood of protests from critics about global warming projections, what we are really hearing is a cry of pain. The pain is real, too. To attack global warming we must profoundly change how we generate and use energy. Producing energy without carbon dioxide emissions can be done, even now, using nuclear fission sources, as France and Japan have done. But fission-generated energy is not safe. Trading one environmental fiasco for another is not an option. We must make a dash to develop better solutions quickly enough to get them into operation and still head off further planet-wide problems of the greenhouse effect. We must repledge our commitment to the Earth and replenish the resources needed for our righteous mission, the quest for the scientific Grail—to develop fusion energy while at the same time accelerating the development of alternative energy sources. Indeed, there are many alternatives—like solar- and wind-generated energy, like hydrogen and photovoltaics and fuel cells—that we should get on-line quickly and utilize widely. This is not the time for ideal solutions alone. It is time for all safe, practical, short-term solutions to be pressed into action while we rush to develop the ones we need for the long-term. And we must rush.

If the change from fossil fuels is not done wisely, particularly if it is postponed until the emergency is in its full-blown stages, we will have an energy crisis as well as a climatic one.

The last time we went through an extended period of energy crisis was in the 1970s, after the Yom Kippur War. Those were the days of a deep freeze in Oregon and mounting disorder in the Midwest, where the U.S. National Guard was stationed on overpasses to prevent violence by independent truckers. Men and women who made their living using diesel oil suddenly found they could not fulfill the contracts they had signed. Energy not only heats and lights our lives but it moves everything. Transportation is the heartbeat of an economy. The '70s were times of prolonged economic stagnation and inflation. This combination mystified economists, but not physicists. They knew that if the price of energy went up, so would everything we called "precious." As a result, we would have to buy less, and, since economies run on sales volume, economies would slow down. The economies of the developed countries were nearly strangled while those of the oil-producing countries surged, at least temporarily, and those of many poorer countries, those with nothing, collapsed. So, here in the '90s, this is the dark task we face—the one of ending fossil fuel use without new sources in place, of trying to feed 6 billion people on a planet that, without fossil fuel, might feed 2 billion.

However, it may still be possible to shut down coal- and oil-fired power plants if we can develop safe, alternative, concentrated energy sources that can produce electricity and utilize our current power distribution infrastructure. Arguably, this is the scientific and technological task we should have undertaken two decades ago. We actually made a good start on the problem in the U.S., until the Reagan administration killed the initiative. However, there is no doubt about it now. This is the critical path and now is the time to act. We can't debate any longer whether the trapped cat is alive or dead. Roll up your sleeves and grab a crowbar. Let's open the canister and rescue the cat while it's still breathing.

If we are to have any hope of avoiding famine and economic chaos, we need a solution which maintains a strong economy, because economic pain is real pain. It puts people out of work, it kills hope, it destroys families and health. Equally damaging, it kills the possibility of commitment to an energy transition. Underfed, underpaid, cold humans will tend to choose short-term comfort over longer-term solutions, even those which might bring a warmer, more prosperous future. So, those who speak glibly of revamping the economy to use drastically less energy are a bit like those who minimize the effects of global warming—they either lack imagination or simply misunderstand the human condition. This is not an argument against conservation and greater efficiencies, both of which will help delay the effects of fossil fuels, create profits for companies and savings for everyone in the

248

process. However, ultimately, as a single strategy, limiting energy use leads to a dead-end when growing populations or expanding economies exceed the net energy benefit from any conservation or energy-efficiency gains. Therefore, unless we can enforce limits on population and growth, as a bottom line we will need a vast new source of power and fuel, and if the majority of our efforts have been placed just on conservation, we may fail to come up with the alternatives we so desperately need.

We should also beware of those who speak glibly of imposing great change on humanity, demanding that we all live up to some new, grand, glorious ideal as a strategy for planetary survival, and we must be careful not to trade our freedoms for a pipe dream offered by those who might resort to fascism if their ideals do not quickly produce the ecological utopia they imagined. So, be wary of those who claim they have the keys to a new paradise and suggest that we can move forward by moving backward to a "simpler time," because driving humanity back to the Stone Age in an effort to stop global warming is an absurd denial of what we could achieve. We don't need to return to a former paradise, we can create a future one—one which embraces both technology and ecology. We only need to adjust our course by turning the wheel toward a new heading.

To stop greenhouse warming and restore Earth, and with it, our balance, requires a new ethic, one that respects both environment and humanity equally. We must view humanity and human economy as intertwined parts of the whole ecology. Can we begin to see human possibility not as an infection, but as the possibility of planetary mutualism, of humans becoming capable partners to an Earth which nourishes all life? Can we see our economy, not as a separate system which aims to dominate Earth's resources, but as a system which could track the managed well-being of people and the planet together?

According to many creation stories, humans once lived in a beautiful garden, where we lived happily until we abused its privileges and fell from grace. In the face of current reality, this story has particular poignancy, as if it belonged, not to a distant past, but to our lives today. To stop greenhouse warming we must act in love toward our neighbors: other humans, plants, animals, trees, and the skies and oceans. This means recognizing the value of

249

humanity and the Earth, and recognizing the frailties of both when contemplating any bold action. Like the Hippocratic oath of medicine which says, "first, do no harm," we must avoid seriously disrupting the human economy and the established ecology when we move to battle global warming. The challenge is this: we must find solutions to greenhouse warming that minimize negative economic effects. Fortunately, when most humans are given a plan that is fair and practical, we are willing to make changes, even those that call for sacrifices, if we know that it will make a significant difference to our friends, family, and children. Many of us will act to save the Earth as well.

To change our basis of energy from fossil fuels requires rational and careful choice. It is useful to briefly consider the sources of energy we already have.

Hydroelectric power comes closest to being ideal. It is concentrated, so simple infrastructures can be used. It makes no greenhouse gases, produces no wastes and no waste heat, so is very efficient. Hydroelectricity uses the power of the Sun to place liquid water at high elevations so it can run downhill. It is really a form of concentrated solar power. However, hydroelectric power means damming rivers, and thus has a high environmental cost and is limited in scope to a few areas.

Next is solar energy. This, too, produces no greenhouse gases and no wastes, but direct solar power is diffuse and must be concentrated or converted to electricity. Solar energy also requires abundant sunlight, so heating London or Moscow in the winter may be highly challenging. Solar power requires quantities of materials that are energy intensive to produce, such as silicon. This leads to the problem that solar plants can, in some instances, cost more in energy to create than they can be expected to generate. To sum up, solar energy is clean but difficult to concentrate efficiently.

Although we have made a great deal of progress in utilizing solar energy, it is a challenge to compensate for the fact that, without substantial concentration, the Sun's rays do not boil water. On the one hand, this is what keeps water liquid over most of this planet, allowing humans to live here in the first place. On the other hand, this means solar energy requires more hardware and surface area to produce the same amount of energy as coal burning, for example. Nevertheless, new techniques, such as using molten salt as a heat collector in a concentrator, may make it especially practical for tropical or semiarid areas where electricity is scarce and sunlight is abundant.

As new materials are developed, such as solar voltaic materials made of plastic and using organic dyes, solar energy becomes ever more useful. Today, advances in photovoltaic roofing materials are making solar energy use more practical for many types of buildings, and it is likely that dramatic new technologic advances in solar energy will continue to be made.

While it may be true, given today's technologies, that solar energy cannot completely replace our current use of fossil fuels, the path of alternative energy is ultimately a great one, full of promise, and the key to perhaps half or more of our future energy needs. Its facets are many and beyond the scope of this book, but generally involve new and expanding applications to harness solar, wind, geothermal and tidal forces. Many of these new inventions hold great promise. Solar voltaic cell technologies have come down dramatically in cost and Japan expects to harvest 4,600 megawatts of electricity by 2010 using newly available solar photovoltaic roofing materials.[1]

For a brilliant look at the possibilities of solar power, we recommend the ongoing work of the Worldwatch Institute. Of particular interest is the chapter "Reinventing the Energy System," in their book *State of the World 1999*.[2] But solar power also has its limitations. The concentrations of energy necessary to meet the needs of huge metropolitan areas, or the energy-intensive agriculture necessary to feed the planet's billions, may well be beyond the capacity of solar power.

Use of wind to generate electricity expanded by over 25 percent in 1998 and renewable energy sources provided over 19 percent of the world's total energy needs.[3] We are getting closer to harnessing tidal and geothermal power, too.

Coal is the worst of the fossil fuels used to make energy. As a "solid to gas" reaction, it is difficult to make its burning "complete." Coal is mostly carbon with lots of impurities that form either air pollutants or toxic solid waste.

251

Because it burns so poorly, coal is both the ultimate greenhouse gas and pollution producer. However, it is cheap and plentiful in many parts of the world.

Natural gas is the best fossil fuel because it is mostly hydrogen and, as a gas, it is easy to purify and burn cleanly. Its chief byproduct is water vapor, followed by carbon dioxide. Burning natural gas is a "gas to gas" reaction that is easy to control. It can burn at high temperature and this means it can create power very efficiently. Natural gas will probably be the most common interim fuel as fossil fuels are phased out. Natural gas is abundant and is often burned off as a waste product in oil fields. Used in fuel cells, it could substantially replace oil and coal on an interim basis for electricity and transportation.

Nuclear energy, in both its fission and fusion forms, has defined our age. To understand the relationship between fission and fusion, we can use the analogy of coal and natural gas. Fission is the "coal burning" of nuclear power and fusion is similar to burning natural gas. Fission works, but it leaves incredibly dangerous waste products that are of such inordinately long-term persistence that they have never been dealt with satisfactorily. Also, fission reactions can be highly unsafe; being vast and fast, they can quickly surge out of control. For example, Chernobyl, a poorly-designed reactor, and Three Mile Island, a well-designed reactor, were both run abominably, and in both cases, short-term oversights led to huge, dangerous, health-damaging, long-term problems.

Fission, the nuclear energy we currently use, is the stored energy of supernovas. Uranium is formed in the heart of supernova explosions, capturing same of its violence for later. So, fission is relatively easy to harness. After the nucleus is split, the energy is released over a span of 10,000 years. The rate of energy released is roughly logarithmic, so most of the energy is released in the first second, with progressively less being released in the next year, and even less in the next century, and so on. As well as making fission unstable, this long period of energy release causes core meltdowns and a high-level radioactive wastes. Additionally, every new nuclear reactor effectively trains a new team in much of the technology of nuclear-bomb building.

As bad as radioactive wastes are, and they are terrible, on a par with them is the carbon dioxide waste formed by fossil fuels, since it can be thought of as having a toxic effect that lasts forever. It is effectively indestructible. Given that both fission and fossil fuels create major long-term problems of toxic wastes, in our opinion, fusion energy is the first choice.

Fusion energy is the "natural gas" of nuclear energy. Fusion produces the energy of a burning star, like our Sun. It makes helium out of lighter hydrogen or helium isotopes. Fusion exists in different reactions. Of these, DT fusion is the easiest to produce, but it makes neutrons in addition to helium and is thus radioactive. The next easiest, D–He$_3$, produces no radioactivity at all. While DT fusion—in its easiest form to harness—does produce low levels of radioactivity, its main energy release is immediate. Meltdowns that head to China aren't a problem for fusion reactors and there is no "high-level" long-lived waste of the type that is the problem with fission. The radioactivity from a fusion plant burning DT will be gone in a few decades, rather than thousands of years, and it will be gone in an hour from a D–He$_3$ plant. Fusion reactions occur, not in a solid-fuel form like fission, but as a diffuse plasma that looks like a flame.

Practical fusion power does not yet exist. It will take an "all-out" development project, including a large investment of funds, to generate the practical process needed to produce usable levels of electric power economically from a fusion reactor. However, fusion is the only power source that will allow greenhouse warming to be stopped with a minimum of economic disruption and environmental costs. It has the potential to be as powerful and limitless as fossil fuels or fission power, yet without long-lived effects or dangers. It is a nuclear power that is much cleaner and safer than fission. While it is neither completely clean nor completely safe, fusion will compare favorably with any other source of large-scale power. So while DT is not perfect, fusion in its more developed forms, such as D–He$_3$, can come very close to our ideal of limitless power at negligible environmental cost and, with continued development, it is getting ever closer to commercial viability.

The schedule for developing fusion technology is not an exact one since it requires progressive experimentation, with each new experiment producing huge amounts of new information, which is then applied to the next round of research. However, in the eminently qualified opinion of noted plasma physicist Dr. Bruno Coppi, a leader in fusion experimentation in both the U.S.

and Europe, and a physics professor at MIT (Massachusetts Institute of Technology), a highly focused effort should be able to produce dramatic results fairly quickly. In his words, it will take "three years to prove the scientific viability of fusion. By then, the scientific basis of fusion will have been proven by producing large amounts, perhaps 100 megawatts or more, of fusion energy."[4] After this initial experimental phase, with the economic viability of fusion power clearly established, it will then take a few more years, perhaps seven, to move it to the marketplace. This timetable is widely accepted within the fusion community.

Dr. Coppi agrees with the authors of this book: What is required to ensure the timely development of fusion energy is a U.S. and European Union–led world-wide initiative to develop it. The timetable for this should reflect a level of urgency similar to the Manhattan Project, which developed the atomic bomb during the Second World War, or the Apollo Program, which put mankind on the moon. Both of these programs were successful because both the scientific community and government agencies were strongly committed to producing a successful result quickly. "It will take more than just throwing money at the problem or funding theoretical studies on fusion," says Dr. Coppi. "It will require coordinated, large-scale experimentation. However, we know exactly what experiments to do and are ready, given the opportunity, to do them."

Of course, unlike the Manhattan Project, which was designed to win a war, the effort here is intended for the benefit of all humanity—to produce the energy needed to electrify the Earth without pollution. The possibility is for a world where, within our lifetimes, all countries have essentially unlimited, cheap, locally-produced electricity. With this, we'll all have the basic requirement for the creation of a world of plenty, where life is secure enough for the population to stabilize and where heat, light, food, and shelter are available to all.

The rose-colored helium plasma glowed in its vacuum vessel as my assistant John Kline and I watched in our research laboratory. Vince

DiPietro (yes, our fellow Mars investigator is also one of the most talented hands-on engineers in the business) skillfully raised the voltage by increments as the capacitors charged. It was a technique he'd discovered: that if the capacitors were charged slowly, arcing would not occur. In the room next door, its lights extinguished to allow us to see the plasma dance and shimmer, stood our experiment—the CMTX or Colliding Micro-Tori Experiment. It was a revival, funded by NASA, of an experiment first carried out by Professor Dan Wells at the University of Miami in the 1970s. Dr. Wells' experiment was said to have achieved fusion conditions on a tabletop. We were now trying to see if it was true.[5]

The CMTX was small, even smaller than Wells' original device. It could fit on a desktop so was quite a contrast to the huge operation at Sandia so many years ago. It created two spinning smoke rings of magnetic field plasma and then fired them into each other, creating a single, larger and hotter smoke ring. Then that in turn was squeezed by pulsed magnetic forces to produce an outburst of amazing heat. All this would happen in a few millionths of a second.

Vince's job had been to construct most of the CMTX device. John Kline, with his usual wizardry, had built the spectrograph that would analyze the flash of light given off by the compressed plasma. I got to worry about what it all meant and, as far as possible, to ascertain whether fusion conditions had actually been achieved. As I anxiously waited and John Kline checked the timing circuits of the fast spectrograph, Vince called out the voltages: 10,000, 12,000 . . . finally reaching 15,000 volts.

"Arm . . . and . . . fire," called out Vince calmly. The lab flashed with brilliant blue-white light as the helium plasma was compressed by a pulse of magnetism. A barely audible "ping" from the Pyrex vacuum vessel was all that could be heard. Our "shots" were going well now, almost like clockwork. The numerous equipment and other problems we'd started with had been solved slowly, one by one. "Safe," he said, as the already discharged capacitors were grounded as a safety precaution.

"Look at this!" exulted John Kline over a beautiful pair of bright spectral lines from the plasma—an indication the "shot" had been hot. We labored on, trying to find a shortcut to fusion, hot fusion on a table top. We finally hit a million degrees in the CMTX before we finished. It was a stunning victory but it was still a temperature found only 60 or so miles into the Sun, a factor of ten below fusion temperatures and not yet near its core temperatures. However, we'd reached this goal both cheaply and elegantly and, given the opportunity, at the next level we'll push on higher and hotter. That is the next step in our journey to achieving workable fusion on a desktop.[6]

W e have been making progress with fusion energy all along, even with only a "back burner" commitment to its development. But with an all-out, well-funded effort, like the one used to put mankind on the Moon, we could have a safe, practical, and economic fusion reactor technology within ten years. We could then begin to replace fossil fuels in powering our electrical systems.

Powering our electrical grids without fossil fuels is only the first major step needed. We also need to use an alternative method of power for our systems of transportation. Using natural gas could act as an interim solution, but what is really needed is a practical electric car. When our vehicles are electrified, they will essentially be using the same energy source that powers the electrical grid. Hydrogen, which when combusted makes water vapor, also holds hope as a fuel. It requires an investment in infrastructure, since pipes and tanks that will hold natural gas will not hold hydrogen, and it can be difficult and costly to store and transport. However, hydrogen is also an excellent non-greenhouse fuel and any real breakthroughs in the development of hydrogen energy should be welcomed since it is a clean and potentially economical source. Fuel-cell technology, which produces electricity from a chemical reaction between hydrogen and oxygen aboard the vehicle, is encouraging, too, as are other newly emerging fuels.[7] Nevertheless, at this point, most of our efforts should be directed toward producing electricity which can be both stored and transported. The infrastructure to distribute electricity already exists in much of the world.

The car should be reinvented and the electric train resurrected, but to do without cars would mean radical economic and social change. Humans love mobility and travel is for many of us our greatest joy and wish. Americans, especially, love mobility and consider it part of their cherished personal freedom. Before the car was even invented, Alexis de Tocqueville, a French visitor to the U.S., noted in his book *Democracy in America* that he had never seen so many horses, and that an American would rather ride his horse around the block than walk. The desire for mobility is no justification for loading the atmosphere with carbon dioxide; it is, however, an indication of the expectations of those in many developed countries. If clean methods of transport can be achieved, people all over the world would no doubt be eager to share the benefits of increased mobility.

Capacity for action is what saves or condemns human beings in crisis. Nothing is required of the helpless. That is why statements that global warming is inevitable and unstoppable are so seductive in the developed world. "If the situation is hopeless, what is the point of doing anything?" we might be tempted to reason, without acknowledging how self-serving that reasoning is. So, hopelessness is a scam with a payoff. There is a great payoff in being relieved of any requirement for action. Those in the First World have their place in the lifeboats after all, and have hands to cover their eyes as the crowded decks become awash. The television cameras turn away from the dying, too. No gain, but no pain either. But it isn't just people who die. Gone too is hope, creativity and, ultimately, salvation.

However, there are those of us who will not be seduced by despair. The attractions of dynamic action are far more appealing: living and breathing, for example. The key, therefore, is to invent rather than lament, to rescue others rather than be claimed as a victim, to break out of the box rather than be buried in it, to seize the day rather than be consumed by it. Humanity and Earth now call for those who "can do." For those who "will do" it is time for decisive action, for accomplishing the near-impossible. The call is out for those who will lead the charge into the future. And if you don't care to lead or follow, don't stand in the way. We can't afford to waste our efforts debating the reality of global warming with those who cover their eyes with one hand and their wallets with the other, when there is abundant evidence for those who are willing to see it.

The stakes of what lies ahead are so high that the decision of each individual counts heavily. In our own lives we've each seen the opportunities to act and save a life or ease misery. We have all seen people pass or fail the tests of these moments. Sometimes a simple action at a critical moment can be what makes a difference for tens of thousands of people. Such a historical moment has arrived for us in the industrialized world. Geography, culture, and the accidents of history have awarded us an abundance of freedom, wealth, and power to take action beyond the wildest dreams of our ancestors, and far beyond the condition of many of our neighbors.

We cannot continue on the path of fire. The fire will consume the air we breathe and its heat will turn our world into a desert. Yet, as long as the financial markets remain unaffected, some may even urge this course. Their priorities could bring disaster to Earth, for to continue mindlessly on the path of fire is to wander aimlessly toward despair. To fail to make decisions, to fail to act decisively says that our only hope is to muddle through. It says we will not save ourselves, even though the correct path through the darkness is fully illuminated and has bright arrows pointing the way. It says we have not the intellect, nor the organization, nor the energy to save ourselves.

Neither can we choose the path of ice. If we shut down our industrial and technological economy to preserve the Earth, many of us would die of cold, hunger, and starvation. We of the First World cannot say to those in the Third World, "You cannot live as we do. You must be content with your simpler life." It is a no-room-for-you-in-the-lifeboats statement.

We cannot turn our backs on technology and live without fire. Our machines pump water and help feed and clothe our children. Technology is essential to our lives. We cannot go back to live in caves and gather herbs, nor chase after some nostalgia-driven utopian ideal. That is the ultimate intellectual fallacy that will lead to mostly empty lifeboats: lifeboats only for the philosophically or spiritually pure. Remember that those who live in caves and gather herbs and hunt must build fires to cook and to keep their children warm and to keep the predators at bay in the night. If we must gather wood to build fires, the cycle only begins again.

There are those who feel that losing nine-tenths of the world's population would be a reasonable strategy to regain balance on Earth, but they offer neither themselves nor their friends as volunteers, leaving the rest of us as the intended candidates. But why would we, even if there were far fewer of us, want to repeat the entire cycle of evolution when we stand poised to break free to some new future possibility? Ultimately, both the path of fire and the path of ice lead to death and despair.

There is another path, however; one between fire and ice. It is the path warmed by dynamic action and innovation and cooled by knowledge, love, and self-restraint. It is the path of technologies that are appropriate for both our Earth and our needs. With the will to discover what appropriate technology is, to invent and develop a future based on both planetary and human needs as well as the bottom line, coupled with the desire to use and share what we invent, we will be able to sustain life here on Earth for millennia to come.

The *Carpathia* arrived 45 minutes after the sinking of the *Titanic*—sadly not soon enough for most of those in the water but just in time for those in the lifeboats who were beginning to die from the cold. The *Carpathia* achieved 17.5 knots in its dash through a field of icebergs to save the *Titanic*, 3 knots above what was supposed to be its maximum speed. It saved 705 lives before daybreak. Captain Rostron and the crew of the *Carpathia* were honored worldwide and enjoyed both the glory of their heroism and the undying gratitude of the survivors and the world at large. To a greater world, their heroic dash rescued both humans and joy from a dark night of folly and death.

As for the other ship, the *Californian*, and its master, Captain Lord? He was fired in disgrace for his failure to respond to the distress flares and didn't sail again until the First World War, when he captained a munitions ship. The *Californian* changed its name because no one would ship cargo on her after her deadly negligence. Although Lord repeatedly tried to clear his name, insisting that the thick ice field would have made any attempt to rescue the *Titanic* impossible—even if he had responded to the flares or turned on his radio—the final judgment was succinct: "An attempt should have been made . . ."

We have given a warning of impending disaster, a disaster we have seen in the distance, on Mars, and all around us on Earth. We have been clearly shown that if this world is carried away by disaster, nature will not pause to shed tears for us. The question boils down to this: on which ship will you serve in these perilous seas? Will it be the ship of cowed inaction disguised as self-indulgent incomprehension, or the ship of bold and skillful action? We propose that we move ahead at full speed and with all our strength and intellect to meet this crisis and avert it—and we welcome you aboard.

Therefore, *carpe diem*. Seize the day! Seize the life of Earth and her inhabitants from the jaws of death! Seize the Sun, too—as solar power in all its forms; and seize fusion energy from its very heart.

A plan is laid before you now. It is a demanding plan, a bold venture, and a visionary quest. It is a plan that some will say is impossible or impractical, or reckless. It will appear to be folly to those who think our present course is a good one, or heresy to those who wish to turn backward. But to those who are concerned and willing to act bravely and sensibly, it will be a great quest, an intoxicating pursuit. The great task we propose is called The Garden Earth Enterprise, or The Enterprise for short.[8] It is not meant to be exhaustive; as inventive committed people we will see many more things that can or should be done. It assumes wise government policies and the bullish response of free markets, because they are part of the plan.

The Garden Earth Enterprise

The world community should commit to accomplish the following tasks by 2010. This effort should include all nations. We need to face this great crisis with the courage and strength of numbers. Together we can take on an apparently insurmountable task and win. With a fundamental commitment to freedom, peace, and human dignity, we respectfully address a world audience, but in particular we plead our case to our fellow Americans, who have been slow to acknowledge the need to turn the ship around and make this great effort. We trust that ultimately everyone will join in, for this is a worldwide task. In the hope that all nations will accept this invitation to act together as the peoples of a planet, we urge them to take the following action.

1. Develop fusion power.

Make the goal to have an economically feasible fusion power plant delivering power into the grid within ten years. Fusion produces cleaner and safer electricity than fission or fossil fuel generation, and it can be workable. The Sun has run for 4.5 billion years on its pure, predictable power. Despite years of anemic funding, hot fusion power has made steady progress. We should fund the development of fusion at $2.5 billion a year and crush the energy problem. The failure to develop fusion energy will create a vacuum for an ugly expansion of fission and fossil fuel energy. To develop fusion energy is to create a renaissance. No other form of power can allow us to use our existing energy infrastructure safely and provide the massive sources of power humanity needs to achieve a decent standard of living with minimum impact on the environment. Petroleum is for plastics. Coal is for storage of carbon underground; leave it there.

2. Develop practical electric cars and reinvigorate electric railroads worldwide.

Strong government economic incentives are needed worldwide to create adequate motivation for automobile companies to move to electric-powered vehicles. We should fund large numbers of electric car development programs to solve the many details necessary to achieve cost-competitive vehicles which are fast, economical, easy to recharge, and stylish. For example, research could be concentrated on developing high-charge, environment-friendly, long-haul batteries or battery-equivalent technologies. Electric trains and public transportation systems should be encouraged as well.

Consumers should demand greenhouse-gas-free vehicles. In the interim, internal combustion cars should be switched to propane or natural gas, or any other fuel which produces less carbon dioxide per mile than gasoline or other greenhouse gases. Consumer response to new gasoline-powered cars should be: "Gas. I'm not buying it anymore." Just as consumers took the lead in protesting against the use of CFCs by not buying propellant spray cans, they can make a substantial difference here. We should pledge to drive our current cars or buy only used cars until viable alternatives are available. We should not buy new cars until they fall into line with standards that will help us achieve the Kyoto Protocol goals and end our dependence on fossil fuels. We can let car companies know we are serious about ending global warming by reducing the sales of new gasoline-burning cars.

3. Pay rainforest nations for their oxygen production.

Swap debt for forests. Most of the rainforest-rich nations have staggering foreign-debt loads. Let the rich nations take repayment of this debt in forest land, to be banked in a worldwide growing "trust fund." The rainforest countries need debt relief and industries for forest products other than lumber as a motivation for keeping trees. We need oxygen and places to soak up some of our excess carbon dioxide. Oxygen is actually our most valuable necessity. When economic models are required to include its true value (and the hidden costs of carbon dioxide) we'll be on the right path. We are all clever people. Work something out!

4. Develop solar energy.

From solar shingles to space-based energy collectors, from geothermal and tidal energy harvesting techniques to wind farms and farming fuels such as methanol, keep expanding the many faces of solar energy collection.

5. Fund a vigorous manned and unmanned space program.

While some complain we can't afford space travel, the exploration of space and its associated technologies may very well have already saved this planet once. Space exploration will save us again if we can mine the Moon for helium-3 to permit radiation-free fusion processes on Earth.[9] We also need to develop the technology to deflect or divert the inevitable—an Earth-impacting asteroid or comet. We must be prepared for such an event. Space exploration and the technological developments that derive from it is an investment in our future and may help us discover the technologies for saving our own atmosphere as well. Space travel is important for the human spirit and mind. As we become citizens of the cosmos, it is only natural that we will become better stewards of our own planet.

6. Ratify the Kyoto Protocol and quit complaining about it!

The Kyoto Protocol, while not perfect, is the single best opportunty we have to work together as the people of the planet to control global warming. It endeavors to be fair and reasonable, while at the same time taking on the difficult challenges involved. The Kyoto Protocol should have the support of all nations. The goal for developed countries of achieving a 7.8 percent reduction of the 1990 levels of greenhouse gases is modest, particularly when held up against the certain costs of increasing global warming in human suffering and environmental and economic devastation. The standards of the Kyoto Protocol will require minimal sacrifice compared to the 60–70 percent reductions we need to ensure that global climate is stabilized.[10] Once we commit to the goal, the creative and financial resources we need to achieve it will become available. We must look to a fossil-fuel free future. The technologies we will develop by 2010 may well make fossil fuel burning seem as archaic as a slide rule is beside a 1Ghz computer, and if rich nations cannot make a small financial sacrifice now, how can nations that have almost nothing be expected to discipline themselves? We must all climb aboard. An important mutualist strategy that First World countries can undertake—in order to give back to the Earth's biosphere as well as to contribute to humanity—is to help develop low-cost, carbon-free energy systems for Third World countries so they won't have to repeat First World mistakes.

7. Move closer to nature and practice a lifestyle less dependent on oil and electricity.

Use public transportation more frequently, lower the setting on the air conditioner and the central heating, get a fan. Quit hiding from the warm air of summer and learn to enjoy the outdoors again, take walks, ride bikes. You will be healthier and the world will be cleaner. A significant amount of material, from bottles to tires, can be recycled. Discover what the materials are and practice recycling. There is great economic benefit in energy efficiency gains for corporations and homeowners alike. Remarkably little investment can give substantial savings. Energy equals money. It is difficult to justify spending more for energy than you have to, so make conservation of energy a standard practice and find out how you could save more, both at home and in your workplace. Join or organize a local action group to lobby your local and national government and monitor local industries. Join an international environmental organization and keep up on what is happening on our planet. Encourage your workplace to be responsible for the environment, to undertake environmental as well as economic objectives, to use environmentally-friendly technologies and to innovate new solutions.

8. Triage environmental problems.

Worldwide, we must adapt the medical model of triage so as to begin to prioritize and manage the multitude of environmental concerns. This way we can develop workable strategies to resolve problems and take action more effectively. For example, we can put more money and a larger effort into certain places in the world if doing so will cut greenhouse gases by 1 percent instead of 0.0001 percent. We must accept stewardship of the planet and take on this responsibility proactively; therefore, we must also set up the structures to enable us to do so. Developing a rating system for standards of risk/need so that levels of urgency can be clearly and fairly communicated is an essential step. Remember, triage was developed under emergency conditions; it is a strategy for saving lives. The more critical the situation is, the more urgently rational approaches for assessment are needed, and they should be designed immediately. Borrow from the economic model, too. Economics has developed a system of environmental distinctions that allow aspects of the environment to be assessed, accounted for and managed. Those distinctions need to be expanded to include the hidden costs to our environment and to throw light on the "dirty deficit." When we can closely link our economic and our environmental well-being, we will see improvements in both.

9. Quadruple the size of the Peace Corps and expand its mission to include a worldwide project to reverse global warming.
Expand the Peace Corps to include volunteers from all over the world. Deploy these volunteers as a task force to stop the spread of the desert. Commit elements of the U.S. Forest Service, the Bureau of Land Management, the Army Corps of Engineers and their international counterparts. Encourage all developed countries to commit governmental and financial resources to the task as well. Get private corporations worldwide involved in supporting reforestation. The front lines are in Africa, South America, Central America, India, and China, but there are other regions, too. Look for places to plant trees—and plant them. Innovate. Involve local communities in reforestation projects. Develop strategies for using trees for a variety of uses other than lumber, but also learn how tree plantations for lumber-producing forest can meet human housing needs without resorting to resorting to the more damaging options of concrete and steel construction. (Trees grow back. Iron ore and gypsum mines don't.) Even plastics can be produced from trees as an alternative to fossil fuels. Plant fruit and nut orchards, and tropical plantations; reforest both public and private lands by creating parks, planting rural tree rows and windbreaks, and building pocket forests in cities. Replant entire watersheds. Create incentives for tree planting and forest management as the "pay back" for help with farming, land and water management in underdeveloped countries. Get creative. Lack of rainfall alone does not make a desert—but coupled with poor land management it does. Plant billions of trees.

We will leave you in the trees, for it is clearly here that Garden Earth will be reborn. Remember the co-existing dimensions of chairs? Let us now consider the dimensions of trees. A tree can be a source of wood, but it is far more important than simply that: it is a life form and a life bringer. It is a solitary beauty, growing tall and silent as the years of life speed by. It is an eater of carbon dioxide and a spreader of carbonic acid into the soil beyond its roots. A tree has economic value, since its fruits can be harvested and it can be cut into lumber or processed into paper. A tree is fuel for fires, a bringer of warmth to the deadly chill of winter, but it is also an immobile candle in the face of ravening fire. Sadly, for some, it is a weed to be burned so feed may be grown for cattle. But a tree is so much more. A tree has esthetic value, for example, when planted for landscaping, or when it stands untouched in a forest. A tree is the progeny of its parents and the parent of its children, so it is a member of a family. It is the result of a chain of evolutionary processes, and can be viewed as a unique genetic code. It may someday become part of a house, but before that it is home to birds and squirrels, a haven where porcupines and raccoon can hide from predators—and perhaps a shaded shelter for the occasional hiker. A tree is a member of a community of trees: a forest. But a forest is also the respiration system of the planet, and the ultimate recycler—God's own perfect conservation system. A forest takes in carbonated air and makes wood and fruit, needles and leaves, and then gives mammals, including orangutans and humans, their ultimate nourishment in the oxygen it exhales.

Trees also speak. Who cannot feel in their bones the wind's song through the trees on a dark and wild night, or hear in its gentle murmur the promise of spring? Trees are speakers of life, telling us, as they flourish, that our stewardship of the land is honorable, and, with their death and absence, save for sawn or fire-blackened stumps, that the Earth is out of balance and, at least in that moment, an environmental holocaust. Places of vital, untouched wilderness—whether they be in North America or the Himalayas—are where we go to remember the deepest wisdom of Earth, a wisdom spoken in tones so low we must wait a long, long time to hear it.

But it is not as simple as only preserving the virgin wilderness (although when possible we must), for trees grow and renew, and, given opportunity and nurture, in a few short years, their abundance will offer a verdant carpet of new seedlings, each striving for light, water and nutrients. This demonstration of how powerfully life returns is a reminder to us that, as trees grow, Earth abides. Growing ever larger for 70 years or even 700, a tree is the ultimate gift to the future—a symbol of life's preeminence.

It is equally significant, however, to understand what trees are not. First, by their growth and survival, they are not another piece of Earth turned to parking lot, housing development or cracked desert. They are not another agribusiness farm tract, where all that was part of the land—bushes and butterflies and flowers and animals large and small—is eliminated, leaving overfertilized, monoculture food crops, doused in pesticides, being mechanically readied to feed Earth's ever-growing numbers of human mouths. A forest grown for timber is not another gypsum or iron ore mine, not another essentially irreparable gash in the Earth, a hole where toxic waters will pool and geese who land there will die. For all but a few years early in its growth cycle, it is also not a place where there is no room for tiny woodland animals, for elk, deer and panthers, for native bushes and grasses, for ground-nesting birds and for birds of prey. Trees are not another place where there is no peace to soothe our overwrought senses, not another noisy city, smoke-filled and suffering, not another sign that humans are too self-interested to ensure their own survival.

So why aren't we growing more trees? Why are we risking the possibility of a catastrophe by putting liquid carbon into the ocean when we could sequester carbon by growing forests? When many trees grow together—many more than are here now, where there are whole forests where now there are only deserts and rough scrub—we will know that Earth and oxygen will survive, and that humanity has accepted its stewardship and finally earned its right to survive as well.

Plant a tree! It is a windbreak and soil stabilizer; it is food for a myriad of microbes and insects; it is a historian of climate, its rings telling us the patterns of past seasons and droughts. A tree is an air-cooling system, and a gentle sponge against flooding. It is the ultimate representative of life on Earth: a literal tree of life.

For some, a tree is sacred. For others, it is a future chair. The essential distinction here—whatever the tree is used for—is one of perception, whether that tree is seen as a "product" or as a living entity, given by life to life, and thus one of life's greatest gifts. Understanding this distinction will transform us from parasites to mutualists: when Earth and its resources are no longer just products, but, rather, gifts to be honored and repaid in equal measure, then both the Earth and humanity will surely survive. In the words of the Hawaiian state motto, "Ua mau ke ea o ka aina i ka pono"[11] (The life of the land will be perpetuated in righteousness).

The Hawaiians understand what we are still learning. Our planet's life depends upon *our* virtue, *our* honor, *our* stewardship and *our* right action; it is a mutual relationship and a worthy enterprise, to be forever and ever perpetuated in love.

Tea With Sophia on the Great Sand Sea

After three days the dust storm abated and it was safe to go above ground. A quick survey revealed that no damage had been done to the facilities so I zapped a quick jump plan to HQ, hopped in a nearby sand-cat and punched in the code to make my escape. I had been cooped up for too long and needed to get out, to rove above ground once again. After speeding through the newly dusted landscape for a dozen miles or more, I pulled off road and continued to cruise, guided only by my need to wander. When all evidence of civilization had long since dropped away behind me, I brought the vehicle to a stop and got out for a walk in the midst of the Great Sand Sea.

Climbing to the top of the nearest. dune, I surveyed the surrounding landscape. As far as the eye could see beautiful rose quartz sand dunes rippled in still waves beneath a cold blue sky flecked with brilliant clouds of ice crystals. The cold sky held no threat, however, since the dark-blue jumpsuit I was wearing provided adequate warmth. Relieved to be outside again, I took a deep breath of the light air, glad that the oxygen levels were sufficient without the burden of supplementation. For the first time in centuries, NASA's trusty old fusion-driven oxygen machines and the massive forestation projects in the equatorial regions far to the south were producing enough of the "O" stuff.[1]

I was just about to turn back to my vehicle when a spot of movement on the top of a nearby dune caught my eye. It was a cat. Nothing special really— just a regular housecat, *Felis domestica*—although what it was doing there in the middle of nowhere was anyone's guess. The cat, ignoring me, sat down.

Silhouetted darkly against the bright dunes, it began licking its paws and washing its face.

Just as suddenly as the cat appeared, a coaxing voice floated over the pink crystal dunes. "Schrödinger! Here, kitty, kitty," it called.

The cat turned its face toward me, and with an expression of utter annoyance, it stretched and began to walk away, in the opposite direction from where the voice was coming. Just as it vanished over the crest of a dune, a feminine figure, in a pink floral kimono billowing loosely in the breeze over her regulation jumpsuit, appeared just where the cat had previously lain. She looked like a butterfly.

"Have you seen my cat?" she inquired, raking her loose windblown hair from her face as if to get a better look at me. "John?" she asked, tilting her head and staring intently now. "Is that you, John Brandenburg?" She marched right toward me and something about her determined stride seemed very familiar.

"Sophia!"[2] I exclaimed as we fell into a warm embrace like the two old friends we were. "What are you doing here in the middle of nowhere?" I laughed aloud when I realized the absurdity of our meeting here on the arctic Martian frontier. Sophia laughed too, her joy expressed by her generous smile. At one time Sophia's polyracial good looks would have seemed quite exotic to me, but her bronzed complexion and almond-green eyes were prototypical Earthling these days. How humans have changed, I thought.

"Come. Tell me what you're doing these days," she suggested with a turn and a mysterious smile over the shoulder that would have induced me to follow her anywhere. "Forget the cat. Let's have tea!"

We followed the tracks she'd left in the sand back to a neat encampment—nothing more than a small solar tent, an airlift shell for supplies and a small round table with two camp chairs. The table, as unlikely as this seemed, was draped with a lace tablecloth, and there, in the middle of the wilderness of Mars, sat a china teapot with a steaming spout, a silver tray burdened with small cakes and all the appointments for a proper tea service.[3] "Please, have a seat," Sophia suggested graciously while extending her hand, her words bringing me back to my senses. "To what do I owe the honor of this unexpected pleasure?" She placed a china cup and saucer in front of me, and pushing back the flowing sleeve of her kimono, she gracefully poured tea.

"We'll . . . ah . . . I really didn't know you were here," I began.

"Well, of course not!" she offered. "No one knows that I'm here. It's a vacation, John. Ever heard of one?" she asked rhetorically, knowing my tendency for overwork. "But since Schrödinger has abandoned me, I'm very glad you came," she smiled.

I had a sudden intuition that maybe this meeting wasn't as accidental as it seemed. After all, there were two chairs. Must be getting suspicious in my old age, I concluded. But this was Sophia after all . . . "I guess we're both pretty busy," I sighed, suddenly relaxing and feeling my weariness, "what with the success of the Institute for Planetary Remediation and all. It's been a long time, hasn't it?"

"Ages," she smirked, knowing full well that it was no exaggeration. "Who would have dreamed it would all turn out so well? We were worried that humans would become extinct, but look at them now! They're everywhere. You can't get away, even on Mars. Nothing personal . . ."

As usual, she was toying with me. "Frankly, it's your own fault, Sophia. We could have had Mars all to ourselves, but you had to go and make Earth a good example for other planets and now property values for the whole solar system are through the roof."

"My fault? I don't think so. It took billions of people to turn it around. Blame them. We had a lot of help." Sophia grew reflective as she sipped her hot tea. "The Enterprise really was the first time humanity aligned on a project like that, wasn't it? Remember when it started?" she asked.

Like I could ever forget. "What a mess!" I offered. "It took years of staggering effort and courage to really come to terms with the problems, sort them out and design workable strategies to resolve them, but when we all realized what was at stake, and we saw it could be done fairly, everyone started to cooperate."

"That, and the fact that business–as–usual was impossible." Sophia sighed.

"Well, Earth is doing just fine," I acknowledged, lifting my tea cup and meeting Sophia's in a toast. "So well, in fact, that it's crawling with tourists from other worlds these days." I smiled, knowing that Sophia would get the joke. "In fact, I'm back there next month."

"To Earth?" she asked, surprised at my revelation. "Why?"

"Oh, they still manage to lure me back there from time to time. Next month I'm meeting a delegation from Alpha Centauri," I explained.

"They're having environmental problems?" she asked, suddenly serious.

"Well, I'm not sure. You remember how it is. Lots of blustering about natural cycles and factions blaming each other before getting down to brass tacks," I suggested.

"Almost like humans . . ." She smiled and squeezed my hand.

"Yes, almost that bad," I smiled and squeezed back.

The Sun set quickly in the thin atmosphere and the glow of Moonrise gave way to the huge silver battered orb of Phobos rising swiftly above the southern horizon. Soon Diemos, its little sister, followed it.

"I'm really going to miss this place," I told her, feeling a chill as the temperature dropped. I found myself wondering if those solar tents really did work.

"Oh, I've a feeling you'll be back," she said, taking my hand and guiding me toward the tent's small silver dome. "Besides, your ocean isn't finished yet."

As Sophia knelt to pick up the mewing cat at her feet, I looked to the night sky one last time before stepping inside. There, suspended in the inky ether, I saw our beautiful Earth, a brilliant blue-white, like Sirius in color but more brilliant than Venus. It had risen in the south over the vast sands of Mars. With a heart full of gratitude that all was well in the cosmos, I said a silent prayer of thanks.

Appendix A

OID as a Global Concern: Magnitude, Causes, and Remedies

by John E. Brandenburg, Ph.D.,
Monica Rix Paxson, and Stephen Kelley Corrick

OID (Oxygen Inventory Depletion), the decrease in atmospheric oxygen concentration, can be predicted based on a simple model that considers only the dominant terms in the global source-sink equation for oxygen. The chief source is the global photosynthetic capacity P, and the sink term is primarily global combustion and respiration S. Thus we have for an approximate equation for atmospheric oxygen concentration O over time, $O,t = P-S$.[1] Since the terms for CO_2 sources and sinks are approximately equal to and negative to the terms for oxygen, this equation is approximately the negative of that for CO_2. Therefore, it is possible to predict the magnitude of Global OID approximately by the magnitude of CO_2 increase, which is approximately 50 parts per million (ppm) since 1958.[2] Thus the present oxygen concentration should be approximately 209,426ppm, down from the 209,476ppm of the 1976 standard atmosphere[3] (which used 1958 measurements). Such a decrease in global oxygen would not be a threat to human health, but would be yet another indication that human activities are out of balance with the environment. Simply put, OID is caused by the human race consuming oxygen faster than it can be replenished. This consumption, primarily for energy production purposes, occurs at the same time as humanity is reducing the global photosynthetic capacity by deforestation and pollution of the oceans. Remedies for OID would include aggressive programs of reforestation and curtailment of oceanic pollution, as well as major government investment in the development of non-fossil fuel energy sources such as hydrogen, solar, and fusion. (See Chapter 9, note 14.)

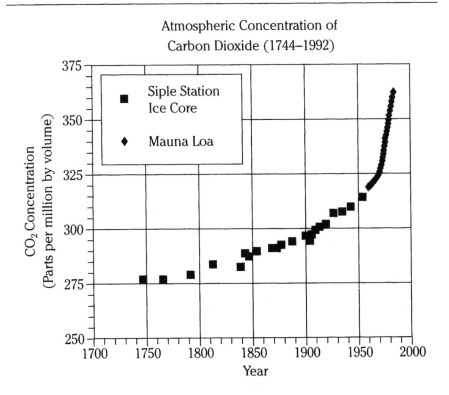

Atmospheric Concentration of
Carbon Dioxide (1744–1992)

The 1998 atmospheric carbon dioxide increase of 2.88 parts per million was the greatest one-year increase in recorded history.[4] (Mauna Loa Chart courtesy of Dr. Michael Pidwirney.)

Appendix B

Peak Flow Meter Test

You are invited to be part of an experiment—to do a little science with us and help us collect some preliminary data. Here is what we are looking into.

There have been increasing levels of carbon dioxide in our environment, particularly the atmosphere, and this trend is predicted to continue unless we radically alter our use of fossil fuels. We thought it would be a good idea to see how carbon dioxide affects our bodies, particularly our breathing.

If you like, you can join in a simple preliminary test, unless for some reason it is unsafe for you to drink carbonated beverages (for example, in some people with asthma this will trigger an attack). If drinking carbonated beverages is fine for you, then you can proceed with the experiment, for which you will need two things.

The first is a "peak flow meter," a device which measures "peak expiratory flow rate"—the fastest speed at which you can blow air out of your lungs after taking as big a breath as possible. It is a very simple plastic device that is used by people with asthma to monitor their own breathing. Since asthma is so common these days, peak flow meters are fairly easy to obtain. Try your local pharmacy or medical supply store. There are online sources as well.[1] A very inexpensive one is fine. Better yet, borrow a peak flow meter from someone you know with asthma.

The second thing you will need is a standard 16 fluid ounce can of seltzer water or a 50 centiliter bottle of carbonated water. We suggest using seltzer water because it has no sugar or caffeine in it (which are drugs) and typically contains no artificial colors, sweeteners, flavors or dyes which might alter the test results. The beverage we want you to drink should be water and carbon dioxide only. Check the label.

What will happen to your peak flow (breathing) when you do the test? We don't know exactly. That's why we're doing it. It's your body and your future, so we figure you just might want to know too.

The Peak Flow Meter Test

1. Review "How To Use A Peak Flow Meter" steps 1–4 (see below). Steps 6 and 7 do not apply to this test.
2. Check your peak flow (write down the results).
3. Drink a 16oz carbonated beverage. (Seltzer water or carbonated water are best.) It should be
   - non alcoholic
   - caffine free
   - sugar and sweetner free
   - have no artificial color, dyes, flavors or sweetners
4. Check your peak flow immediately (write down the results).
5. 10 minutes later, check your peak flow (write down the results)
6. 20 minutes later, check your peak flow (write down the results).
7. E-mail the results to the Peakflow Group (peakflow@egroups.com).

If you are willing to share your peak meter, get your friends and family members to take the test, too.

Thank you for your participation.

How to Use a Peak Flow Meter
(Instructions issued by the American Lung Association)

Step 1. Before each use, make sure the sliding marker or arrow on the Peak Flow Meter is at the bottom of the numbered scale (zero or the lowest number on the scale).

Step 2. Stand up straight. Remove gum or any food from your mouth. Take a deep breath (as deep as you can). Put the mouthpiece of the Peak Flow Meter into your mouth. Close your lips tightly around the mouthpiece. Be sure to keep your tongue away from the mouthpiece. Blow out as hard and as quickly as possible. Blow a "fast hard blast" rather than "slowly blowing" until you have emptied out nearly all of the air from your lungs.

Step 3. The force of the air coming out of your lungs causes the marker to move along the numbered scale. Note the number on a piece of paper.

Step 4. Repeat the entire routine three times. (You know you have done the routine correctly when the numbers from all three tries are very close together.)

Step 5. Record the highest of the three ratings. Do not calculate an average. This is very important. You can't breathe out too much when using your Peak Flow Meter but you can breathe out too little. Record your highest reading.

Step 6. Measure your Peak Flow Rate close to the same time each day. You and your doctor can determine the best times. One suggestion is to measure your Peak Flow Rate twice daily between 7 and 9 a.m. and between 6 and 8 p.m. You may want to measure your Peak Flow Rate before or after using your medicine. Some people measure Peak Flow both before and after taking medication. Try to do it the same way each time.

Step 7. Keep a chart of your Peak Flow Rates. Discuss the readings with your doctor.

© 1999 American Lung Association. Reprinted with permission.

Notes

Foreword

1 "Climate Change and Greenhouse Gases," policy adopted in December 1998 by the Council of the American Geophysical Unions <http://www.agu.org/sci_soc/policy/climate_change.html>.

Chapter 1

1 See Appendix A.

2 Ralph Keeling, Stepher Shertz, "Seasonal and interannual variation in atmospheric oxygen and implication for the global carbon cycle," *Nature*, vol. 358, August 27, 1992, p. 723–7. In this study OID is described as actually three times the rate we predicted. However, the data is given in such an arcane formulation that no lay person would recognize it. Also, Wallace Broecker, Jeffrey Severinghaus, "Diminishing Oxygen," *Nature*, vol. 358, August 27, 1992, pp. 310–11.

3 Interview, Michael Rose, 1998.

4 "A great daylight fireball over the Northwest," *Sky and Telescope*, October 1972, pp. 269–72.

5 Ibid.

6 "Fireballs flash," *Colorado Springs Gazette Telegraph*, August 11, 1972, p. 1.

7 "Meteor shower flames across West," *The Sun*, August 11, 1972, p. 1.

8 Luigi G. Jacchia, "A meteorite that missed the Earth," *Sky and Telescope*, July 1974, pp. 4–10.

9 *Colorado Springs Gazette Telegraph*, ibid.

10 Jack London, "To Build a Fire," *Lost Face*, The Macmillan Company, New York, 1910.

11 "Asthma—United States, 1980–1990, Centers for Disease Control," <http://www.cdc.gov/epo/mmwr/preview/mmwrhtml/00017708.htm>.

12 Audit—Hospital Admissions for Asthma, St. John's Hospital website, <http://www.equip.ac.uk/maag13/hospital.html>.

13 American Petroleum Insitute, *Oil & Gas Journal*, December 1995.

14 "Caspian oil flow starts in April, Azerbaijan says," *Chicago Tribune*, October 27, 1998.

15 Stephen M. Brown, et al., "The Effect on mortality of the 1974 fuel crisis," *Nature*, vol. 257, September 25, 1975, pp. 306–7.

16 Mark Hertsgaard, "Our real China problem," *The Atlantic Monthly*, November 1997.

17 "Twenty percent of Korea's air pollutants blown from China," *Asian Meteorology Online Newsletter*, <http://rossby.metr.ou.edu/~spark/AMON/v1_n5/News/China.html>.

18 "Asian smog seen affecting U.S. air quality," Reuters, December 6, 1998.

19 Hertsgaard, op. cit.

Chapter 2

1 The ecliptic plane is the orbit of Earth as projected on another celestial body, in this case, Mars.

2 Martin Caidin, Jay Barbree, Susan Wright, *Destination Mars in Art, Myth, and Science*, Penguin, New York, 1997, p. 87.

3 David Fisher, *Fire and Ice: The Greenhouse Effect, Ozone Depletion and Nuclear Winter*, Harper & Row, New York, 1990, p. 59.

4 Ibid, pp. 61–2.

5 Joseph Husband, *A Year in a Coalmine*, Houghton Mifflin, Boston and New York, 1911.

6 American Lung Association, Asthma Statistics.

7 Barry E. Digregorio, Gilbert V. Levin, Patrica Ann Straat, *Mars: The Living Planet*, Frog, Berkeley, Calif., 1997, pp. 140–1.

8 Radio interview with Gilbert Levin and Ron Levin, Laura Lee Show, July, 1997, archived on Realtime Audio.

9 Just as this book goes to press it would appear that Ron and Gilbert Levin's contention that Mars has a blue sky has been supported by photos from the Hubble telescope. What NASA is calling "true-color" images show that Mars clearly has more Earth-tones than usually depicted in other images of the planet. The sky appears as a blue haze. "Hubble Image Descriptions," Florida Today Space Online, February 26, 1999, <http://www. flatoday.com/space/explore/probes/hubble/1999/022699a.htm>.

10 Stanley V. McDaniel, *The McDaniel Report: On the Failure of Executive, Congressional and Scientific Responsibility in Investigating Possible Evidence of Artificial Structures on the Surface of Mars in the Setting Mission Priorities for NASA's Mars Exploration Program*, North Atlantic Books, Berkeley, Calif., 1993, p. 25.

11 Gilbert V. Levin, Patricia Straat, "A Reappraisal of Life on Mars," Proceedings of the NASA May Conference, July 21–23, 1986, vol. 71, Science and Technology Series, University of California, San Diego, p. 187.

12 "Mauna Loa Summit Observatory" online article located at <http://mloserv.mlo.hawaii.gov/history/mlosummi.htm>.

13 Fred Reedy, "Propane-fuel car runs well; fleet will follow," *Tribune Herald*, Hilo, Hawaii, March 15, 1972.

14 "Car won't dirty up Mauna Loa air," *Tribune Herald*, Hilo, Hawaii, March 14, 1972.

15 See CO_2 graph from Mauna Loa in Appendix A.

16 Reedy, op. cit.

17 "Our Changing Planet: The FY 1998 U.S. Global Change Research Program," A Report by the Subcommittee on Global Change Research, Committee on Environment and Natural Resources of the National Science and Technology Council, a Supplement to the President's Fiscal Year 1998 Budget.

18 Policy Implications of Greenhouse Warming—Synthesis Panel, Committee on Science, Engineering, and Public Policy, National Academy of Sciences, National Academy of Engineering, Institute of Medicine, National Academy Press, Washington, D.C., 1991.

19 Intergovernmental Panel on Climate Change (IPCC), 1996, Summary for Policymakers, *Climate Change 1995—The Science of Climate Change: Contribution of Working Group I to the Second Assessment*, Cambridge University Press, U.K.

20 "Climate Change and Greenhouse Gases," policy adopted by the Council of the American Geophysical Union in December 1998, <http://www.agu.org/sci_soc/policy/climate_change.html>.

Chapter 3

1 David C. Knight, *Robert Koch: Father of Bacteriology*, Granklin Watts, New York, 1961, pp. 39, 41.

2 *Biographical Dictionary of Scientists, Biologists* (General Editor Peter Abbott), Bedrick Books, New York, 1983, p. 77–8.

3 Dov Trietsch, Thoughtlets, 1997, <http: www.mgsltns.com/koch.htm>.

4 Related to meme (pronounced "meem"): A meme (term coined by Richard Dawkins, by analogy with "gene") is a contagious information pattern that replicates by parasitically infesting human minds and altering their behavior, causing them to propagate the pattern. Individual slogans, catchphrases, melodies, icons, inventions, and fashions are typical memes. An idea or information pattern is not a meme until it causes someone to replicate it, to repeat it to someone else.

5 David E. Nye, *Electrifying America: social meanings of a new technology, 1880–1940*, MIT Press, Boston, 1995.

6 Jim Motavalli, "Power struggle: will utility deregulaton finally unplug 'dirty' electricty?", *The Environmental Magazine*, November/December 1997, <http://www.emagazine.com>.

7 "Solar architect Steven J. Strong says that Reagan's policy of slashing funding for renewables cut the industry off at its knees. 'I'm very bitter about what Reagan did to the solar industry,' he says. 'I saw a lot of my friends lose their businesses, go bankrupt, lose their homes. It was not just a lack of enthusiasm for renewables. It was a deliberate, vindictive and orchestrated campaign to snuff it out in as forceful and as vehement a manner as they could.' Strong has a point—the official neglect killed a growing industry. In 1980, the U.S. had 233 solar collector manufacturers, shipping 19,398 units; by 1992, there were only 45 manufacturers, shipping 7,000 units." Tracey C. Rembert, "As Americans worry about rising oil prices, global warming and a disintegrating nuclear industry, renewable energy is making a dramatic comeback," *The Environmental Magazine*, November/December 1997.

8 *World Resources 1998–1999*, a joint publication of The World Resources Insitute, The United Nations Environment Programme, The United Nations Development Programme and The World Bank, Oxford University Press, U.K., 1998.

9 Lester R. Brown, Michael Renner, Christopher Flavin, *Vital Signs 1998: The Environmental Trends that are Shaping Our Future*, for the Worldwatch Institute, W. W. Norton, New York.

10 Interview with Ralph Nader, *Rolling Stone*, November 20, 1975.

11 Michael Rogers, "Spray away layers of ugly ozone fast, fast, fast!" *Rolling Stone*, December 4, 1975.

12 Barry E. DiGregorio, Gilbert V. Levin, Patricia Ann Straat, *Mars: The Living Planet*, Frog, Berkeley, Calif., 1997, p. 257.

13 John Dworetzky (editor), *Introduction to Child Development*, 2nd, West Publishing, 1981.

14 Starburst Pixel Interleaving Technique (SPIT) was the first of many new developments in imaging technology that started with the investigation of the image of the "face" on the surface of Mars 20 years ago. To obtain proof that this technique was not distorting the image, the SPIT process was tried on some Landsat (Earth satellite) data of Dulles Airport. The SPIT-processed image revealed extraordinary detail that was not apparent in the original. So the technique worked well. It was so good in fact that Goddard Space Flight Center gave Vince DiPietro and Greg Molanaar an award for developing the algorithm.

Chapter 4

1 Ann Marie Low, *Dust Bowl Diary*, University of Nebraska Press, 1984, p. 95.

2 *The American Heritage History of the 1920s and 1930s*, (Editor in Charge, Ralph K. Andrist), American Heritage/Bonanza Books, New York, 1988, p. 305.

3 Ann Marie Low, op. cit., p. 95.

4 Arthur M. Schlesinger Jr., *The Coming of the New Deal*, Houghton Mifflin, Boston, 1959, pp. 69–70.

5 Ann Marie Low, op. cit., p. 95.

6 Carl Sagan, *Cosmos*, Random House, New York, 1980, p. 108: "We could wander down the ancient river valleys, up the slopes of one of the great volcanic mountains, along the strange stepped terrain of the icy polar terraces, or muster a close approach to the beckoning pyramids of Mars. The largest are 3 kilometers across at the base, and 1 kilometer high—much larger than the pyramids of Sumer, Egypt or Mexico on Earth. They seem eroded and ancient, and are perhaps, only small mountains, sandblasted for ages. But they warrant, I think, a careful look."

7 David E. Fisher, *Fire and Ice: The Greenhouse Effect, Ozone Depletion & Nuclear Winter*, Harper & Row, New York, 1990, pp. 119–28. The scientists who were involved in the real discovery were Carl Sagan, Richard Turco, Brian Toon, Thomas Ackerman, and James Pollack. Thanks for the great work!

8 In Carl Sagan, op. cit., p. 129.

9 James F. Kasting, "The origins of water on earth," *Scientific American*, vol. 9, November 3, 1998, pp. 19–20.

10 Baron Johannes von Buttlar, "Mars— The New Earth," in *The Case for the Face* (eds. S. V. McDaniel and M. R. Paxson), Adventures Unlimited Press, Kempton, Ill., 1998, pp. 209–18.

11 Harry Moore, "Ice in Craters," American Geophysical Union Conference, May 28, 1998.

Chapter 5

1 One of the best sources of helpful information on how humans can go to self-destructive lengths to avoid feeling pain and how to find the way back is Dr. M. Scott Peck's *The Road Less Traveled: A New Psychology of Love, Traditional Values and Spiritual Growth*, Simon & Schuster, New York, 1998.

2 "The Carbon Bomb: Climate Change and the Fate of the Northern Boreal Forests", Greenpeace International, <http://www.greenpeace.org/index.shtml>, 1994. This report is an extremely thorough analysis of the boreal ecosystem under stress.

3 University of Kentucky's Coal Information website at <http://www.uky.edu/KGS/coal/webcoal/pages/coal3.htm>.

4 The second major path is through the ocean carbon cycle and sedimentation.

5 Brian J. Stocks, "The extent and impact of forest fires in northern circumpolar countries," in *Global Biomass Burning: Atmospheric, Climatic and Biospheric Implications* (ed. Joel S. Levine), MIT Press, Cambridge, Mass., 1991.

6 Allan N. D. Auclair and Thomas B. Carter, "Forest wildfires as a recent source of CO_2 at mid-high northern latitudes," *Canadian Journal of Forest Research*, 23, 1993, pp. 1528–36.

7 Olga N. Krankina, "Forest fires in the former Soviet Union: past, present and future greenhouse gas contributions to the atmosphere," in *Proceedings of the International Workshop on Carbon Cycling in Boreal Forest and Sub-Arctic Ecosystems* (eds. Ted S. Vinson and Tatyana P. Kolchugina), U.S. Environmental Protection Agency, Global Change Research Program, 1991.

8 Stocks, op. cit.

9 Greenpeace International, op. cit.

10 N. Myers, "Taiga taiga burning bright," *Guardian*, Manchester, August 5, 1993.

11 Judith Matloff (of the *Christian Science Monitor*), "Raging infernos and illegal logging threaten Siberia," *The Sunday Oregonian*, October 18, 1998.

12 One source of acrimony did appear, however, when I discovered, much to my anger, that Richard Hoagland had changed our abstract after I had submitted it. He had added some discussion about his "city," a feature he found significant. I felt that in taking the lead on the paper I had done so at enormous professional risk, and the abstract I had written was the maximum I considered defensible. I felt betrayed. In retrospect, although the changes introduced by Hoagland were minor, any desire to continue working with him after the conference left me.

13 Our party consisted of myself, Vincent DiPietro, Richard Hoagland, and Dr. Thomas Rautenburg.

14 I marveled at how well the paper looked when we laid it out. Vince had prepared most of it and used large computer lettering and many images from Viking to illustrate our points. I still have the impression of a miraculous effort, given the limited resources of time and money that IMIT possessed.

15 Leonard David, "Mars researchers

complain of data lag from MGS," *Space News*, September 7–13, 1998. Mike Malin, head of Malin Space Science Systems (MSSS), a NASA contractor controlling the camera aboard the Mars Global Surveyor (MGS), is considered by many independent Mars researchers to be a hostile opponent capable of delaying, withholding and distorting data to serve his own agendas. Complaints are heard throughout the Mars research community, as this article makes clear.

16 Sagan and Fox, *Icarus*, vol. 33, 1973, p. 72.

17 It is interesting to note that Steve Squyers, with Cornell University, has been investigating the possibility of life on Europa, a moon of Jupiter.

18 J. E. Brandenburg, *The PaleoOcean of Mars*, MECA symposium on Mars: Evolution of its Climate Atmosphere, Lunar and Planetary Institute, Houston, Technical Report 87–01, 1986, pp. 96–8. Although this is the first scientific paper published on the Mars PaleoOcean, the idea of an ocean on Mars was first introduced by David L. Chandler in his book, *Life on Mars*, Clarke, Irwin and Co. Ltd, Toronto, 1979.

Chapter 6

1 Mark C. Knutson, "The Ponzi Scheme," 1996 <http://www.usinternet.com/users/mcknutson/pscheme.htm>.

2 While in this case we are talking about the environment, it is interesting to note that the number of financial Ponzi schemes has also increased tremendously over the last few years. The US Securities and Exchange Commission (a governmental agency created in the 1930s after the 1929 stock market collapse) has had an increase in the number of complaints filed. One of the most dramatic failed Ponzi schemes in recent history (it collapsed in 1997) involved thousands of Albanians in a scheme where hundreds of millions of dollars were lost. The President of Albania, Sali Berisha, explained that his government's failure to intervene in the swindle was a reflection of its commitment to the free market. How do you distinquish a Ponzi scheme, which probably looks for all the world like a real opportunity to participate in American style capitalism, to those with so little experience of the real possibilities of the market place?

3 E. Schrödinger (1935), "The Present Situation in Quantum Mechanics," *Naturwiss*, 23, pp. 807–12, 825–8, 844–9. "One can even construct quite burlesque cases. A cat is shut up in a steel chamber, together with the following diabolical apparatus (which one must keep out of the clutches of the cat): in a Geiger tube there is a tiny mass of radioactive substance, so little that in the course of an hour *perhaps* one atom of it disintegrates, but also with equal probability not even one; if it does happen the counter responds and through a relay activates a hammer that shatters a little flask of prussic acid. If one has left this entire system to itself for an hour, then one will say to oneself that the cat is still living, if in that time no atom has disintegrated. The first atomic disintegration would have poisoned it. The ψ-function of the entire system would express this situation by having the living and the dead cat mixed or smeared out (pardon the expression) in equal parts." "It is typical of such cases that an uncertainty originally restricted to the atomic domain has become transformed into a macroscopic uncertainty, which can then be resolved through direct observation. This inhibits us from accepting in a naive way a 'blurred model' as an image of reality . . . There is a difference between a shaky or not sharply focused photograph and a photograph of clouds and fogbanks."

4 Buzz Aldrin, *Return to Earth*, Random House, New York, 1973.

5 Russian Research Center "Kurchatov Institute" Hypertext Database: Chernobyl and its consequences (Project "Polyn") Chernobyl Global Radiation Patterns, <http://webnuc.nuce.psu.edu/~chernoby/glbrad.html>.

6 Chernobyl Status, April 1997,

<http://www.greenpeace.org/~comms/9
7/nuclear/reactor/chern11.html>.

7 Ibid.

Chapter 7

1 Barry E. DiGregorio, Gilbert V. Levin,
Patricia Ann Straat, *Mars: The Living
Planet*, Frog, Berkeley, Calif., 1997, p. 260.

2 At least that was the official story. In
reality, in one major test the Viking mission
had produced positive results that might
have indicated life on the surface of Mars,
but it had been summarily dismissed.

3 DiGregorio, et al., op. cit., p. 260.

4 Op. cit.

5 J. E. Brandenburg, "Constraints on the
Martian Cratering Rate Based on the SNC
Meteorites and Implications for Mars
Climate History," *Earth, Moon and
Planets*, vol. 67, 1995, pp. 35–45.

6 Support for Brandenburg's hypothesis
of the Martian cratering rate as four times
the lunar rate is found in L. E. Nyquist, L.
E. Borg, C. Y. Shih, "Shergottite age
paradox and the relative probabilities of
Martian meteorites of differing ages,"
Geophysical Research Journal (Planets),
vol. 103, no. E13, December 25, 1998, pp.
31–445.

7 Robert P. Crease, Charles C. Mann,
Timothy Ferris, *The Second Creation:
Makers of the Revolution in Twentieth-
Century Physics*, Rutgers University Press,
Piscataway, N. J. 1996, p. 44.

8 Ibid., p. 162.

9 J. E. Brandenburg and V. DiPietro, "Did
the Lyot Impact End the Liquid Water Era
on Mars?", American Geophysical Union
Conference, Baltimore, June 1995, pp.
52A–3.

10 DiGregorio, et al., op. cit. p. 289.

11 Ibid.

12 BioTech Life Science Dictionary at
<http://biotech.chem.indiana.edu/>.

13 Peter W. Price, *Evolutionary Biology of
Parasites*, Princeton University Press,
Princeton, N.J., 1980, p. 11.

14 Ibid.

15 Ralph Keeling, Stepher Shertz,
"Seasonal and Interannual variation in
atmospheric oxygen and implication for
the global cycle," *Nature*, vol. 358, August
27, 1992, pp. 723–7.

16 FrogWeb, "Focus on Amphibian
Declines and Deformities," National
Biological Information Infrastructure,
<http://www.frogweb.gov/>.

17 An extraordinary book on the efforts
of the "brownlash" movement to discredit
scientific environmental findings is
*Betrayal of Science and Reason: How
Anti-Environmental Rhetoric Threatens
Our Future*, by Paul R. Ehrlich and Anne
H. Ehrlich, Island Press, Washington, D.C.,
1996.

18 Kate Dourian, "Britian says world
cannot afford inaction on climate,"
Reuters, November 5, 1998.

19 "The woes of too much CO_2," ABC
News, May 22, 1998.

20 Karen Schmidt, "Coming to Grips
With the World's Greenhouse Gases,"
Science, vol. 281, July 24, 1998, p.
504a–506a.

21 P. D. Quay, B. Tilbrook and C. S.
Wong, "Oceanic uptake of fossil fuel CO_2:
carbon-13 evidence," *Science*, vol. 256,
April 3, 1992, p. 74.

22 Jonathan Broder, "Miracles in tragedy
of Cameroon," *Chicago Tribune*, August
30, 1986.

23 Robert Cooke, "Disarming deadly
waters," *Newsday*, Nassau and Suffolk
edition, May 21, 1996, p. 821.
Volcanologist Haraldur Sigurdsson from
the University of Rhode Island was the
first to suggest that gases in the lake were
involved. William Evans, research chemist
at the U.S. Geological Survery in Menlo
Park, California, joined him in the request
for funding.

24 There is a very interesting discussion of
the underlying tectonics of Cameroon at
<http://volcano.und.nodak.edu/vwdocs/fr
equent_question/grp7/africa/question627.
html>. Most significant is the opinion that

Cameroon, rather than lying on a hotspot trace, is perhaps a leaky transform fault from the mid-Atlantic ridge.

25 Christie Morrison, "The killer lakes of Cameroon: strange and deadly events in Cameroon crater lakes instigate inquiry and raise concern for the future", <http://www.ent.msu.edu/~long/nyos.html>.

26 "Scientists to draw gas from lake to prevent disaster: Cameroon," *The Gazette*, Montreal, March 26, 1995, p. 85.

27 Morrison, op. cit.

28 Ibid.

29 Death by soda pop? While it would seem that soda pop is harmless, those in the business of handling carbon dioxide for soda must regard the substance with respect, as this letter posted on OSHA's website, at <http://www.osha-slc.gov/OshDoc/Interp_data/I19960528.html>, would indicate.

OSHA Standards Interpretation and Compliance Letters

5/28/1996—The Dangers of Carbon Dioxide Exposure

January 9, 1996

Honorable David L. Hobson

150 N. Limestone St.

Room 220

Springfield, Ohio 45501

Dear Representative Dave Hobson,

In the early morning of September 26, 1995 a 58-year-old male delivery driver collapsed and died shortly after arrival at Greene Memorial Hospital. The investigation revealed the driver had delivered Carbon Dioxide from his semi-tractor trailer to a local fast food restaurant and was found approximately 20 minutes later at the bottom of a darkened stairwell. The driver was found sitting upright in the corner of the stairwell, still wearing his face shield required by OSHA in delivery of carbon dioxide. The autopsy revealed no apparent reason for death with the

exception of the blood work drawn at the hospital that disclosed an extremely high level of carbon dioxide in his blood at the time of death. I ruled the death an accident resulting from carbon dioxide intoxication. I discovered the carbon dioxide system utilized is a common system used to operate the carbonated drink dispensers in numerous restaurants and businesses. High concentrations of carbon dioxide, which displaces oxygen, can result in death in less than 15 minutes.

I would like your assistance in alerting those that work with or come in contact with carbon dioxide, of the dangers associated with exposure to the carbon dioxide gas. Having completed the medical aspects of this investigation, I am making the following suggestions and recommendations for public and private workers' safety. These recommendations are for prevention of further accidents or fatalities involving Carbon Dioxide Intoxication.

1. All new Carbon Dioxide receptacles are installed at ground level in an open area.

2. The remaining underground receptacles have an adequate ventilation system installed.

3. If any detection system for Carbon Dioxide is available, it be installed near the receptacle to alert workers if the concentration reaches a dangerous level.

4. Periodic inspections and, if needed, replacement of the delivery and receiving system in all of the facilities that utilize a detrimental gas distribution system.

5. General safety inspections to replace inoperable lights, and to correct identified safety hazards.

The dangers involved with Carbon Monoxide (CO) have been well documented and publicized. Some publicity has been afforded to Carbon Dioxide (CO_2) dangers in confined spaces associated with agricultural settings and underground settings where gas can accumulate naturally. This death and associated dangers don't apply to the agricultural or confined space hazards, as the stairwell wasn't enclosed and the driver introduced (delivered) the Carbon Dioxide to the atmosphere. Although this death may appear isolated, this is a common carbon dioxide system utilized to operate carbonated drink dispensers, and there is potential for further accidents and fatalities. With your assistance maybe we can alert workers and the general public of the dangers associated with exposure to concentrations of Carbon Dioxide. If you can use these recommendations or have others and I can be of assistance to help promote them, please let me know.

Sincerely,

M. R. Desai, M.D.

Greene County Coroner

Chapter 8

1 William McDonough, Michael Braungart, "The NEXT Industrial Revolution," *The Atlantic Monthly*, October 1998.

2 Bill Blakemore (ABC News correspondent), "A personal reflection on ecology, wars and peace: the fate of Forests in Lebanon and Israel," ABC News and Starwave, Corp., 1998.

3 It is interesting to note that the issue of Pluto's status as a planet was raised by scientists in January of 1999 as well. "Pluto may lose planet designation," Associated Press, January 20, 1999.

4 In *Journal of Scientific Exploration*, vol. 5, no. 1, 1991.

5 J. E. Brandenburg, V. DiPietro, "Support for the New Mars Synthesis Derived from MGS (Mars Global Surveyor) and Other Data," American Geophysical Union, Spring Meeting, June, 1999. *EOS Supplement* p. S209, April 27, 1999. Also, the new Mars Orbital Laser Altimeter data suggest that Cydonia resided near the lowest land on Mars at the deepest part of the atmospheric gravity well, and, therefore, at the ideal elevation for high temperatures, a dense atmosphere and life. See Laboratory for Terrestrial Physics

<http://ltpwww.gsfc.nasa.gov/tharsis/mola.html> for a beautiful topographical map of the elevation differences between Mars' Northern and Southern hemispheres.

6 William K. Hartmann, "Cratering in the Solar System," *Scientific American*, vol. 236, January 1977, pp. 84–6.

7 Scott McCartney (Associated Press), "Defense labs sharing their secrets", *Chicago Tribuness* May 6, 1990.

8 "Iraqi troop buildup triggers U.S. show of might," *Chicago Tribune*, July 25, 1990.

9 Asthma is not the only disease which has environmental causes. The United States Environmental Protection Agency reports that there has been a steady increase for two decades in new cases for some cancers among children—60 percent in testicular cancer, 50 percent in one form of bone cancer (osteogenic sarcoma), 30 percent in brain cancer (glioma), and 10 percent in one form of blood cancer (acute lymphoblastic leukemia). This agency feels that environmental factors may play a part in these cancers. Conference to Address the Increase in Childhood Cancer announcement, Environmental Protection Agency, September 11, 1997. This certainly represents a step in a positive direction in solving this problem.

10 Brigid Schulte (Knight-Ridder Washington Bureau), "Sperm counts on steep decline in U.S., Europe," *Lexington Herald-Leader*, November 24, 1997, <http://www.kentuckyconnect.com/heraldleader/news/112497/ffsp.html>.

11 Gordon K. Durnil, *The Making of a Conservative Environmentalist*, Indiana University Press, Bloomington, Ind., 1995, p. 49 (from a speech given before the Indiana Wildlife Federation annual meeting).

12 I. B. Cohen, "Florence Nightingale," *Scientific American*, vol. 250, March 1984, pp. 128–36.

13 Sally Lipsey, *Newsletter of the Association for Women in Mathematics*, vol. 23, no. 4, July/August 1993, pp. 11–12.

Chapter 9

1 Malcolm W. Browne, "Carbonated water may have caused mass extinction 250 million years ago," *The New York Times*, July 30, 1996, <http://users.owt.com/km_beck/public_html/mas s-e~1.htm>. The scientists who published the report on which this article was based are Dr. Andrew H. Knoll of Harvard University's Botanical Museum, Dr. Richard K. Bambach, a geologist at Virginia Polytechnic Institute and State University in Blackburg, Dr. Donald E. Canfield of the Max Planck Institute for Marine Microbiology in Bremen, Germany, and Dr. John P. Grotzinger, a planetary scientist at the Massachusetts Institute of Technology. See also *Science*, vol. 273, July 26, 1996, vol. 273, pp. 452 ff.

2 Yukio Isozaki, "Permo-Triassic superanoxia and stratified superocean: records from lost deep sea ocean," *Science*, vol. 276, April 11, 1997, p. 237. As an explanation of the mass extinction which occurred at the end of the Permian era... "A model of overturn of CO_2-saturated deep anoxic water paired with a hypercapnia hypothesis appears promising because it explains various aspects of the end-Permian extinction including the remarkable selectivity to organisms and isotopic signatures."

3 There are competing theories about this extinction, chiefly that it also could have been caused by excess volcanic activity, oceanic anoxia (reduced oxygen) or a sea level fall.

4 G. Marland, B. Andrus, and C. Johnston, "Global CO_2 emissions from fossil-fuel burning, cement manufacture, and gas flaring: 1751–1996," Carbon Dioxide Information Analysis Center, Oakridge, Tenn., <http://cdiac.esd.ornl.gov/ftp/ndp030/global96.ems>.

5 Ibid.

6 Wallace S. Broecker, "Thermohaline circulation, the Achilles heel of our climate system: will man-made CO_2 upset the current balance?", *Science*, vol . 278, November 28, 1997, p. 1582–8. "China's rapid industrialization has led to upward revision of predictions regarding the magnitude of this buildup. While previously we thought in terms of doubling the strength of the CO_2 content of the preindustrial atmosphere, current thought is moving toward a tripling."

7 Matthew Jones, "Interview: Coal courts the developing world," Reuters, December 14, 1998. In a recent interview, Ron Knapp, chief executive of the World Coal Institute, based in London, explained, "They will use coal in the way we in developed countries used coal 100–150 years ago—use it as a building block of economic development. There is an enormous demand for energy."

8 Jocelyn Kaiser, "Possibly vast greenhouse gas sponge ignites controversy," *Science*, vol. 282, October 16, 1998, pp. 386–7. As an example, one recent study suggests that North America absorbs 1.7 petagrams (1.7 billion metric tons) of carbon dioxide a year, while another study calculates that North America absorbed only about 0.6 billion petagrams per year between 1988 and 1992. The difference between the two studies' conclusions represents over 15 percent of the total carbon dioxide released by human activities into the atmosphere last year.

9 Jason Webb, "World temperatures could jump suddenly," Reuters, November 4, 1998, See also J. Severinghaus, et al., "Timing of abrupt climate change at the end of the Younger Dryas interval from thermally fractionated gases in polar ice," *Nature*, vol. 391, January 8, 1998, pp. 141–6.

10 Wallace S. Broecker, et al., "A Salt Oscillator in the Glacial Ocean?" *Paleoceanography*, vol. 5, 1990, pp. 469–77.

11 J. M. Barnola, D. Raynaud, C. Lorius, and Y. S. Korotkevich, "Trends: a compendium of data on global change," Carbon Dioxide Information Analysis Center, Oak Ridge, Tenn., 1994.

12 J. L. Sarmiento and J. C. Orr, "Three-dimensional simulations of the impact of

the southern ocean nutrients depletion on atmospheric CO_2 and ocean chemistry," *Limnology & Oceanography*, 36 (8), pp. 1928–50.

13 Ibid.

14 J. E. Brandenburg, M. R. Paxson, S. K. Corrick, "OID (Oxygen Inventory Depletion) as a global concern: magnitude, causes, and remedies," *EOS* (supplement), Transactions, AGU, vol. 79, November 10, 1998, p. F172.

15 The "Death by Soda Pop" hypothesis was suggested by Monica Rix Paxson and Stephen Kelley Corrick with the assistance of John Brandenburg.

16 Browne, op. cit.

17 Wallace S. Broecker, et al., "The biggest chill," *Natural History*, November 1987, vol. 91, pp. 74–82. The particular mechanism which could shut down the conveyor belt is that North Atlantic water, because of heavy evaporation rates, is saltier, and therefore heavier than regular ocean water. This heavier weight causes the North Atlantic water to sink to the bottom and then to begin its long journey back to the tropics.

18 Broecker, "Thermohaline circulation," op. cit., p. 1585. If the Conveyor fails, it is also possible Europe could plunge into a frozen state similar to Canada's Northern Territories, since the belt brings heat north and the huge oceanic carbon sink presently driven by the conveyor would stop, too.

19 U. Mikolajewicz, B. D. Santer, and E. Maier-Reimer, "Ocean response to greenhouse warming," *Nature*, vol. 345, June 14, 1990, pp. 589–98, and, "Changes in ocean currents could accelerate global warming," *New Scientist*, July 14, 1990.

20 A unanimous warning from the executive Board of the American Geophysical Union states: "There is no known geological precedent for the transfer of carbon from the Earth's crust to atmospheric carbon dioxide, in quantities comparable to the burning of fossil fuels, without simultaneous changes in other parts of the carbon cycle and climate system. This close coupling between atmospheric carbon dioxide and climate suggests that a change in one would in all likelihood be accompanied by a change in the other." Therefore, they conclude, "AGU believes that the present level of scientific uncertainty does not justify inaction in the mitigation of human-induced climate change and/or adaptation to it."

21 Lucien Dujardin, "Catastrophe Teacher: an introduction for experimentalists," <http://perso.wanadoo.fr/l.d.v.dujardin/ct/eng-index.html>.

22 Jocelyn Kaiser, op. cit., pp. 386–7.

23 R. F. Keeling, S. Piper and M. Heimann, "Global and hemispheric CO_2 sinks deduced from changes in atmospheric CO_2 concentration," *Nature*, vol. 381, May 16, 1996, pp. 218–21.

24 J. Fan, et al., "A Large Terrestrial Carbon Sink in North America Implied by Atmospheric and Oceanic Carbon Dioxide Data and Models," *Science*, vol. 282, October 16, 1998, pp. 442 ff.

25 *Tellus*, 1999, cited by J. Kaiser, in "Possible vast greenhouse gas sponge ignites controversy," *Science*, vol. 282, October 16, 1998, pp. 386–7. This study (led by P. Monash) calculates that the North American carbon sink between 1988 and 1992 absorbed only about 600 million tons of CO_2, only about 35 percent of what the Sarmiento study calculated.

26 "The Missing Sink" hypothesis was developed by Stephen Kelley Corrick with the assistance of Monica Rix Paxson and John E. Brandenburg.

27 Gene E. Likens, F. H. Brown, and N. M. Johnson, *Environment*, vol. 14, no. 2, March 1972, p. 33; and E. Barrett, and G. Brodin, "The acidity of Scandanavian precipitation," *Tellus*, vol. 7, 1955, pp. 251–7.

28 M. C. Bowles, and H. D. Livingston, "Assessing the ocean's role in the global carbon cycle: The Joint Global Ocean Flux Study (JGOFS)," update of *McGraw-*

Hill's Yearbook of Science and Technology, 1997. Found online at <http://www1.whoi.edu/overview.html>.

29 Increased U.S. and Arctic rainfall. Land-Atmosphere-Ice Interactions Science Management Office, "Land-Atmosphere-Ice Interactions: A Plan for Action," Report to the National Science Foundation, Office of Polar Programmes. University of Alaska, Fairbanks, 1997, p. 50.

30 See the very helpful information on the SWOOPE (Students Watching Over Our Planet's Environment) website at <http://www.wvu.edu/~ruralnet/pswoope/acid/basic.htm>.

31 Joseph B. Verrengia, "Scientists: Rain Spoils Weekend Fun," Associated Press, August 6, 1998.

32 SWOOPE website (see note 30).

33 (AMAP) Arctic Monitoring and Assessment Program, "Arctic Pollution Issues: A State of Environment Report," Oslo 1997. Also (BESIS) "Bering Sea Impact Study: The Impacts of Global Climate Change in the Bering Sea Region," September, 1996.

34 William Krabill, et al., *Science*, vol. 283, March 5, 1999, pp. 1522–4. Also, Michael Kahn, "Shrinking Greenland glacier signals global warming," Reuters, March 4, 1999. The news gets worse and worse. A study published in *Science* says new NASA satellite research concludes that the Greenland glacier has shrunk by over 3 feet since 1993. The implications again are that global warming is accelerating and that there could be significant loss of the total volume of the glacier and of its ability to cool and regulate climate by the middle of the 21st century. If the glaciers of Greenland and Antarctica do shrink dramatically, it would also lead to large rises in global sea levels and a resultant flooding of coastal areas.

35 Myron Arms, *Riddle of the Ice*, p. 128. This is a wonderful, readable book which offers a beautifully written story of a sailing ship's journey north to study the puzzling dynamics of the great influx of sea ice, which has occurred during the last 20 years into the East Canadian/West Greenland cold spot, and of the tremendous work of a number of scientists to try and understand its dynamics.

36 Additionally, this infusion of fresh water into salinated water will tend to mask carbon absorption, since oceanic absorption calculations are based on average ocean salinity and carbon levels. Since carbon is heavier than water, could it then be that this unprecedented infusion of carbon, plunging to the ocean floor as a part of the Great Conveyor Belt's thermohaline cycle, also serves to increase both the speed and volume of the general convection current of the Great Convection Belt? And, the more speed and volume of the resultant downward convection current, the greater will be the final pressures exerted on the concentrations of the carbonic acid on the floor of the ocean. Potentially, the greater the pressure, the greater the actual amount of carbonates which may tend to be fixed in deep suspension in the deeper recesses and valleys of the ocean floor. Remember that, as in Cameroon, in some bodies of water, when cold carbonated water is covered by a layer of warmer water, the temperature gradient acts as a lid to prevent the release of almost all the carbon, so it will passively remain on the floor of the body of water in which it is secreted in a thermal stratum.

37 It is important to note that, while the Northern Arctic is heating up, the area around Newfoundland is actually colder than in the past because of the massive floes of ice now streaming south. Unfortunately, if global warming continues unabated, this ice flow will stop within the next century as the Arctic ice cap mostly finishes melting. When it does, there will be no more temperature differential to drive the Great Conveyor Belt, and, in a stagnant or reversed-current situation, the possibility of massive release of carbon dioxide becomes a deadly threat.

38 Paul G. Falkowski, R. T. Barber, V.

Smetacek, "Biogeochemical controls and feedbacks on ocean primary production," *Science*, vol. 281, July 10, 1998, p. 202.

39 "Report of the Status of Groundfish Stocks in the Canadian Northwest Atlantic," Atlantic Stock Assessment Secretariat, Department of Fisheries and Oceans, June, 1994.

40 Myron Arms, op. cit., pp. 77–8. In a related note, it is disturbing to note that every cessation of a planetary cooling trend in the last 160,000 years has been marked by a massive flow of ice from the Arctic into the Atlantic region between Newfoundland and Greenland (ibid., p. 155). See also, Richard Alley, "Resolved: the Arctic controls global climate change," *Arctic Oceanography*, vol. 49, 1995, pp. 263–83.

41 Jocelyn Kaiser, op. cit.

42 Verrengia, op cit.

43 "U.S. National Report to International Union of Geodesy and Geophysics," 1991–1994, *Rev. Geophys.*, vol. 33 supplement, American Geophysical Union, <http://www.agu.org/revgeophys/schime01/node14.html>.

44 "Fossil Energy Techline," United States Department of Energy, "U.S., Japan, Norway sign first Kyoto Agreement; will jointly sponsor tests for long-term CO_2 disposal," <http://www.fe.doe.gov/techline/tl_co2seq.html>.

45 Malcolm W. Browne, "Carbonated water may have caused mass extinction 250 million years ago," *The New York Times*, July 30, 1996: "Marine animals with calcium carbonate shells or skeletons build them from carbon dioxide and dissolved calcium compounds. But they are particularly sensitive to overdoses of carbon dioxide, called hyercapnia." So, the very presence of too much carbon dioxide in the ocean threatens to kill one of the most important mechanisms for its removal.

46 Why are we risking the possibility of a catastrophe by putting liquid carbon into the ocean when we can sequester carbon by growing forests at a low cost—$3 to $22 per ton, depending on the specific land-use policy? R. D. Perlack, R. L. Grahm, and A. M. Prasad, "Land use management and carbon sequestering in SubSaharan Africa," *Journal of Environmental Systems*, 22 (3), pp. 199–210.

47 Dr. Clark R. Chapman, "Statement on the Threat of Impact by Near-Earth Asteroids," U.S. Congressional Hearings on Near-Earth Objects and Planetary Defense, May 21, 1998, <http://impact.arc.nasa.gov/congress/1998_may/chapman.html>.

48 Exodus 21:29: "If the ox has been accustomed to gore in the past, and its owner has been warned but has not restrained it, and it kills a man or a woman, the ox shall be stoned, and its owner also shall be put to death," *The New Oxford Annotated Bible*, Oxford Univeristy Press, 1994. Much of our present personal injury law can be seen as being derived from this injunction.

49 Cindy Schreuder, "1995 disaster helps give meaning to heat deaths," *Chicago Tribune*, April 2, 1997.

50 For a wonderful discussion of chairs and how our limited thinking about them has physically damaged many of us, see Dr. Galen Cranz's *The Chair: Rethinking Culture, Body, and Design*, W. W. Norton, New York, 1998.

51 One of the classic definitions of economics is the study of the distribution of scarce resources. Those resources are exclusively environmental resources. Even the contributions of humans cannot be thought of as separate from the environment, since humans are part of the natural world and totally dependent on environmental resources.

52 Dr. Janet Yellen, Chair, White House Council of Economic Advisers, "Statement of the Economics of the Kyoto Protocol before the Committee on Agriculture, Nutrition, and Forestry, U.S. Senate," Washington, D.C., March 5, 1998.

53 Allan D. Brunner, "El Niño and World

Primary Commodity Prices: Warm Water or Hot Air," International Finance Discussion Papers, No. 608, Board of Governors of the Federal Reserve System, April 1998, <http://www.bog.frb.ged.us>.

54 The twin terrors of Iniki and Andrew hit within four weeks of one another and took both Homestead, Florida, and the island of Kauai, Hawaii, back to the Stone Age for about a week, eliminating almost all power sources, all sources of fresh water and most normal food distribution supplies. In Florida, the area was in almost total anarchy with rioting and looting rampant until the National Guard was able to make its way into the damaged areas two days after the storm. Source: Earth Systems and Global Change, lecture by Dr. Rick Behl, California State University, Long Beach.

55 From the transcript of the video, "Climate Change and the Insurance Industry," Greenpeace USA, 1993.

56 Greenpeace Climate Impact Database <http://www.greenpeace.org/~climate/database/records/zgpz0706.html>, Seventh Billion Dollar Natural Catastrophe in Three Years.

57 J. K. Leggett, "Climate Change and the Insurance Industry," *Greenpeace International Special Report*, May, 1992. (website as in note 56).

58 Donna Abu-Nasr, "Weather caused $89 billion in damage in 1998," Associated Press, November 28, 1998.

59 Thomas R. Knutson, R. E. Tuleya, Yoshio Kurihara, "Simulated increase of hurricane intensities in a CO_2-warmed climate," *Science*, vol. 279, February 13, 1998, pp. 1018 ff.

Chapter 10

1 J. E. Brandenburg, "Mars as the parent body of the CI carbonaceous chondrites," *Geophysical Research Letters*, vol. 23, no. 9, May 1, 1996. Also, J. E. Brandenburg, "The CI as the Missing Old Meteorites of Mars," Lunar and Planetary Science Conference, Houston, Texas, November 1998. J. Farquaher, et al., "D17O

Measurements of Carbonate from ALH84001: Implication for Oxygen Cycling Between the Atmosphere, Hydrosphere, and Pedosphere of Mars," Lunar and Planetary Science Conference, Houston, Texas, November 1998.

2 Bartholomew Nagy, *Carbonaceous Meteorites*, Elsevier Scientific Publishing, New York, 1975.

3 David Whitehouse, "Sensational new claims about life on Mars are about to be made by U.S. scientists," BBC News Online Science, March 2, 1999, and "Unusual features found on a Martian meteorite may be those of fossilized alien bacteria after all, research suggests," BBC News Online Science, February 11, 1999. Brandenburg's theory of life in the CIs was further reinforced by the announcement by David McKay's team at the Lunar Planetary Sciences Conference in March, 1999, that a second SNC meteorite from Mars, the Nakhla meteorite, also contained fossilized remains of life. Researchers have confirmed that Earth microbes can produce fossil casings similar to those found in ALH84001, the Martian meteorite which first indicated life on Mars. The existence of similar fossils on Earth makes it far likelier that the cylindrical casings found in the Mars meteorite are, in fact, the remains of a living organism from Mars.

4 David Harry Grinspooon, *Venus Revealed: A New Look Below the Clouds of our Mysterious Twin Planet*, Helix Books, Addison-Wesley, Reading, Mass., 1997.

5 Nagy, op. cit.

6 James W. Head, M. Kreslavsky, H. Hiesinger, M. Ivanov, S. Pratt, B. Thomson, "Large Standing Bodies of Water in the Past History of Mars: Further Test for Their Presence Using Mars Orbital Laser Altimeter (MOLA) Data," American Geophysical Union, Spring Conference, 1999, *EOS Supplement*, p. S202, April 27, 1999.

7 Conversation with geologist James Erjavec, a long-time Mars researcher. Also, see his paper "Evidence for a Paleo-Ocean Shoreline, Sedimentary Features and Water Erosion in Cydonia Mensae," American Geophysical Union, Spring Conference, 1999, *EOS Supplement*, p. S210, April 27, 1999.

8 J. E. Brandenburg, and V. DiPietro, "Did the Lyot Impact End the Liquid Water Era on Mars?" American Geophysical Union Conference, Baltimore, MD, June 1995, pp. 52–3.

9 "On Monday, March 30, 1998, Stanley McDaniel and Mark Carlotto, as representatives from The Society for Planetary SETI Research (SPSR), met at JPL with MGS Project Manager, Glenn Cunningham, and MGS Chief Scientist, Arden Albee to discuss the release of images. Because of the controversy surrounding the Cydonia landforms, Cunningham said that NASA would not make any official statements concerning the images when they were released, leaving interpretation to the scientific community. The SPSR agrees with this position." From the SPSR press release, "After Meeting with NASA, Mars Scientists Prepare to Analyze New Images of Cydonia Face," April 2, 1998.

10 The second person was JPL's Glenn Cunningham, who in January 1999 was appointed to head up all Mars missions through to 2014. To his credit, and with the deep appreciation of the Mars Cydonia research community, he has kept his promise and remains publicly silent on any conclusions about the new Cydonia images.

11 Carl Sagan, *Pale Blue Dot*, Random House, New York, 1994, pp. 227–8.

12 "Dry ice blanket kept Mars water flowing," Reuters, December 10, 1997. "We found that this dry ice 'blanket' actually warms the planet because it reflects infrared light back to the surface more than it reflects solar radiation outward," said University of Chicago professor Raymond Pierrehumbert in a statement.

13 "Major Air Pollutants," American Lung Association.

14 Wayne State University School of Medicine, Carbon Monoxide HQ, <http://www.phymac.med.wayne.edu/FacultyProfile/penney/COHQ/Nastygases/CO2nastygas2.htm>.

15 Ibid.

16 Yandell Henderson Ph.D, "Carbon Dioxide," *Cyclopedia of Medicine*, 1940.

17 S. Boljevic, A. H. Kogan, S. V. Gracev, S. V. Jelisejeva, I. G. Daniljak, "Carbon dioxide inhibits the generation of active forms of oxygen in human and animal cells and the significance of the phenomenon in biology and medicine," *Vojnosanit Pregl*, July/August 1996, 53(4), pp. 261–74.

18 I. Ellingsen, G. Sydens, A. Hauge, J. A. Zwart, K. Liestol, G. Nicolaysen, "CO_2 sensitivity in humans breathing 1 or 2 percent CO_2 in air," *Acta Physiol. Scand.*, February 1987; pp. 129(2), pp. 195–202.

19 K. Prowse, C. Duvivier, R. Peslin, R. Sadoul, "The respiratory response to inhaled carbon dioxide in man after 3 hours' exposure to 3 percent carbon dioxide," *Clin. Sci. Mol. Med.*, September 1978, 55(3), pp. 309–16.

20 Wallace S. Broecker, "Thermohaline circulation, the Achilles heel of our climate system: will man-made CO_2 upset the current balance?, *Science*, vol. 278, November 28, 1998, p. 1586.

21 J. Wenzel, N. Luks, G. Plath, D. Wilke, R. Gerzer, "The influence of CO_2 in a space-like environment: study design," *Aviation, Space and Environmental Medicine*, March 1998, 69(3), pp. 285–90.

22 See Appendix B, Peak Flow Test.

23 Our rate of conversion for carbon dioxide from weight to volume is: a cubic foot of 100 percent carbon dioxide weighs 0.1144 pound.

24 Lester R. Brown, Michael Renner, Christopher Flavin, *Vital Signs 1998: The Environmental Trends that are Shaping Our Future*, for the Worldwatch Institute, W.W. Norton, New York.

25 This is the OSHA maximum for short-term exposure.

26 Tracey C. Rembert, "As Americans Worry About Rising Oil Prices, Global Warming and a Disintegrating Nuclear Industry, Renewable Energy is Making a Dramatic Comeback," Earth Action Network, November 15, 1998, <http://www.emagazine.com>. According to this source, the per capita levels of carbon dioxide release in China and India are a mere 1/500th of those generated by the U.S.

27 *The Case for the Face: Scientists Examine the Evidence for Alien Artifacts on Mars* (eds. Stanley V. McDaniel, Monica Rix Paxson), Adventures Unlimited, Kempton, Ill., 1998.

28 Horace W. Crater, President of the SPSR, "The MGS Cydonia Images: Preliminary Report." This report, released July 25, 1998, was prepared for NASA on behalf of the SPSR Research Team. The full report is available at <http://www.mcdanielreport.com/nasarpt.htm>. See also, Stanley V. McDaniel, "Cydonian Mound Geometry," and Horace Crater, and C. Sirvent, "The Beauty of Martian Mound Geometry," both in S. V. McDaniel and M. R. Paxson, *The Case for the Face*. Also, Horace Crater, S. V. McDaniel, M. Carlotto, "Analysis of Landforms in Cydonia Mensae." American Geophysical Union, Spring Confernece, 1999, *EOS Supplementary*, p. S210, April 27, 1999.

29 Mark Carlotto, "Analysis of Global Surveyor Imagery of the Face on Mars," American Geophysical Union Conference, May 28, 1998. Carlotto's work can be viewed at <http://www.psrw.com/~markc/marshome.html>.

30 Harry Moore, "Ice in Craters," American Geophysical Union Conference, May 28, 1998. Harry Moore, J. Brandenburg, S. Corrick, A. Sirisena, "Ice found in Carters in Cydonia," American Geophysical Union, Spring Conference, 1999, *EOS Supplement*, p. S211, April 27, 1999. On July 7, 1999, a new, better image of the Moore crater was taken by JPL and Malin Space Systems. It can be seen at:

<http://barsoom.msss.com/mars/global_surveyor/camera/images/7_8_9_cydonia/index.html>.

31 James Erjavec, "Evidence of Water and Sedimentary Deposits in and Around Cydonia," American Geophysical Union Conference, May 28, 1998.

32 Harry Moore's finding of crater-bound water ice was supported by a Malin poster session at the AGU, which showed an image of a crater on the southern plains of Mars which has evidence of dried-up water rather than the extant water ice of Moore's finding.

33 John E. Brandenburg, Vincent DiPietro, "The New Mars Synthesis and the Cydonian Hypothesis: Models Confront New Data," American Geophysical Union Conference, May 28, 1998.

34 Vince was justifiably outraged by this display and lodged a formal protest with NASA, but the letter in response said that the individual involved "was not a NASA employee" so nothing could be done. Ironically, this same person had acted as the NASA spokesperson to the international media with regard to the MGS images with no such disclaimer.

Chapter 11

1 "Turkey rejects Syrian water-sharing offer," *The Washington Post*, October 16, 1998.

2 "Global warming to kill tropical forest—UK experts," Reuters, November 2, 1998.

3 P. Ciais, P. P. Tans, J. W. C. White, M. Trolier, R. J. Francey, J. A. Berry, D. R. Randall, P. J. Sellers, J. G. Collatz, and D. S. Schimel, "Partitioning of ocean and land uptake of CO_2 as inferred by 13C [carbon-13] measurements from the NOAA Climate Monitoring and Diagnostics Laboratory Global Air Sampling Network," *Journal of Geophysical Research*, vol. 100 (D3), 1995, pp. 5051–70.

4 After reports that the Brazilian rainforest was being destroyed at a rate that had increased by 30 percent in 1998—5000 soccer fields of jungle every day—the Brazilian Environmental Minister announced a review of clearing permits. William Schomberg, "Brazil suspends issuing of Amazon clearing permits," Reuters, November 2, 1999.

5 This prediction was borne out by England's Hadley Centre for Climate Change's November 1998 computer analysis, which projected that, after 2050, global warming may accelerate to the much-feared runaway greenhouse scenario, and that the Amazon, southern Europe and central Africa are all likely to effectively become deserts.

6 "Global warming could grow," Reuters, March 1, 1999, <http://www.abcnews.com/sections/scie nce/Daily News/warming990225.html>.

7 There are a number of published studies which provide additional support for the existence of OID: Ralph Keeling, Stepher Shertz, "Seasonal and interannual variation in atmospheric oxygen and implication for the global carbon cycle," *Nature*, vol. 358, August 27, 1992, p. 354; Michael Bender, "Atmosphere—keynote perspective—carbon cycle studies based on the distribution of O^2 in air," *Tellus*, vol. 51, no. 2, 1999, p. 165.

8 See Appendix B for information on the use of peak flow meters.

9 Centers for Disease Control, "Asthma Mortality and Hospitalization Among Children and Young Adults, 1980–1993," MMRW; May 3, 1996, 45(17), pp. 350–3.

10 "City life ups asthma risk in black children," Reuters, May 4, 1998, <http://www.pathfinder.com/penews/dr weil/indexStory7.html>.

11 "Earth's temperature shot skyward in 1998," *Science News*, vol. 155, January 2, 1999.

12 Greenpeace Climate Impacts Database, <http://www.greenpeace.org/~climate/da tabase/records/zgpz0600.html>.

13 N. Salafsky, "Drought in the rainforest: effects of the 1991 El Niño–southern oscillation event on a rural community in West Kalimantan, Indonesia," *Climatic*

Change, vol. 27, pp. 373–96, August 1994.

14 "Fire and brimstone: Sander Thoenes reports on Suharto's economic woes," *Financial Times*, London, October 8, 1997.

15 "Fires destroy 30 million cubic meters of Indonesian forests per year," Agence France Presse, July 29, 1997.

16 "More than 600 forest fires in Indonesia," Deutsche Presse-Agentur, August 6, 1997.

17 "A burning issue," *The Times*, London, September 17, 1997.

18 "Logging companies' greed feeding Indonesia's forest fires," Agence France Presse, September 25, 1997.

19 "New Zealanders help fuel Indonesian forest fires," Greenpeace, October 28, 1997, <http://www.greenpeace.org/ ~comms/97/forest/press/october28. html>.

20 "Corruption fans the flames as forests burn; Personal and political greed lie behind the blazes in Indonesia," *The Independent*, London, March 15, 1998.

21 "Forest fires reach doorstep of tropical paradise, Bali," *Korea Times*, October 1, 1997.

22 "Deadly fires plague Southeast Asia," *USA Today*, September 24, 1997, <http://rossby.metr.ou.edu/~spark/AMO N/v1_n5/News/elnin/elnino_fire.html>.

23 Greenpeace, 1997, op. cit.

24 Michael S. Serrill, "Catching the Asian Flu," *Time Magazine*, November 3, 1997, vol. 150, no. 18, <http://cgi.pathfinder.com/ time/magazine/1997/dom/971103/nation.ca tching_the_.html>.

25 *Financial Times*, ibid.

26 "Record drought devastates Indonesia," *USA Today*, October 24, 1997.

27 "Rain washes away smog, but causes floods," *USA Today*, November 20, 1997.

28 The $1.3 billion cost for the effects of the Indonesian fire pollution was estimated by the World Wide Fund for Nature's Indonesia program and the

293

Singapore-based Economy and Environment Program for Southeast Asia.

29 "New Indonesian forest fires raise smog concerns," CNN, February 25, 1997, <http://rossby.metr.ou.edu/~spark/AMON/v2_n1/News/fire/smog_IN_CNN.html>.

30 "The fire next time," *The Economist*, London, February 28, 1998.

31 "Smoke shrouds Indonesia's eastern Borneo as fires rage," *Korea Time*, March 22, 1998, <http://rossby.metr.ou.edu/~spark/AMON/v2_n1/News/fire/KT_Indonesia.html>.

32 "The numbers just don't add up," David Liebhold, *Time Magazine*, Asia edition, March 23, 1998, vol. 151, no. 11, <http://cgi.pathfinder.com/time/magazine/1998/int/980323/numbers.html>.

33 "Don't expect a blank check," Christopher Ogden, *Time Magazine*, Asia edition, ibid., <http://cgi,pathfinder.com/time/magazine/1998/int/980323/viewpoint.html>.

34 "U.S. backs $1 billion bailout for Indonesia despite lack of reforms," *New York Times*, News Service, May 1, 1998 Washington.

35 "Forest-fire smoke hits critical level," Tribune News Service, April 12, 1998.

36 "Asia's burning, and the whole world suffers," J. Madeleine Nash, *Time Magazine*, May 4, 1998, vol. 151, no. 17, Notebook/Planet Watch.

37 "The choking drought, southeast Asia is suffering a fiery dry spell unequaled in 50 years, with famine a dangerous possibility," Uli Schmetzer, *Chicago Tribune*, May 2, 1998.

38 "The graves of wrath," *The Economist*, London, May 23, 1998.

39 "Rape widespread in Indonesia," *Chicago Tribune*, December 19, 1998.

40 As note 38.

41 "Asian crisis could spread, Greenspan warns," *Chicago Tribune*, May 21, 1998.

42 "Dead economy has Indonesia staring at prolonged recession," Liz Sly, *Chicago Tribune*, May 29, 1998.

43 "Damage from Indonesian fires, choking haze put at $4.4 billion," *Chicago Tribune*, May 31, 1998.

44 Eugene Linden, "Smoke signals: vast forest fires have scarred the globe, but the worst may be yet to come," *Time Magazine*, June 22, 1998, vol. 151, no. 24.

45 Donna Abu-Nasr, "Weather causes record $89 billion damage," Associated Press, November 28, 1998.

46 "1998 hottest in past 1,000 years: U.S. Researchers," Kyodo News Service, March 4, 1999. Earth's average temperature was 14.46 °C. A research team of meteorologists headed by Michael Mann at the University of Massachusetts and also including researchers at the University of Arizona have concluded that the 900-year downward trend of temperatures was reversed early in the 20th century and has headed upward dramatically in the latter half of the century.

47 Michael Mann, et al., *Geophysical Research Letters*, March 3, 1999.

Chapter 12

1 Lester R. Brown, Michael Renner, Christopher Flavin, *Vital Signs 1998: The Environmental Trends That Are Shaping Our Future*, Worldwatch Institute, W. W. Norton, New York, 1998, p. 18.

2 Lester R. Brown, et al., *State of the World 1999: A Worldwatch Institute Report on Progress Toward a Sustainable Society*, W. W. Norton, New York. The book and the chapter we recommend (written by Christopher Flavin and Seth Dunn) is also available online at <http://www.worldwatch.org>.

3 Ibid, p. 23.

4 From a March 6, 1999 interview with Dr. Coppi. See also, "From Laboratory to Space: Lectures in honor of Bruno Coppi," January 19–20, 1995, <http://rleweb.mit.edu/groups/g-plasym.htm>.

5 J. E. Brandenburg, J. F. Kline, V. R.

DiPietro (Research Support Instruments), "Progress on the CMTX (Colliding Micro-Tori Experiment," a poster session presented at the 1998 Division of Plasma Physics Meeting of the American Physical Society, November 17, 1998.

6 On December 17, 1998, Sandia National Laboratories announced they had achieved temperatures of 2.3 million °C in a machine known as "Z." "Fusion hopeful hits temperature high," *Science News*, vol. 155, January 2, 1999.

7 Another promising, but very new low-emission technology, AquaFuel™, is still in its early testing phases, but seems to burn cleaner than any fossil fuel and can be produced simply by running electricity through recycled carbon electrodes inserted in fresh, salt or even sewer water. AquaFuel™ is a gas process developed by William H. Richardson, Jr. and presently being tested by Toups Technology Licensing, <http://www.toupstech.com/aquafuel>.

8 The Garden Earth Enterprise can be accessed at <http://www.gardenearth.com>.

9 G. L. Kulcinski, E. N. Cameron, J. F. Santarius, I. N. Sviatoslavsky, L. J. Wittenberg and H. H. Schmitt, "Fusion Energy from the Moon for the 21st Century", Lunar Bases and Space Activities of the 21st Century Second Symposium, April 1988, NASA Conference Publication 3166, p. 459. Also, S. W. White, "A Current Bibliography of Helium-3 Research", University of Wisconsin Report UWFDM-1003, January, 1996.

10 From a public address by Ross Gelbspam, author of *The Heat is On: The High-Stakes Battle over Earth's Threatened Climate*, on the subject "Beyond Kyoto: it's time to think big," published in the winter 98 online journal, *e-Amicus: Beyond Kyoto*, <http://www.igc.apc.org/urdc/eamicus/98win/kyoto1.html>.

11 Adopted by Hawaii's first State Legislature, 1959.

Epilogue

1 Discovery Channel Online, "Mars Oxygen Machine Under Way." NASA engineers at Johnson Space Flight Center in Houston have developed an Oxygen Generating System (OGS) which can convert carbon dioxide back to oxygen, <http://www.discovery.online.com>.

2 The name Sophia means wisdom in Greek.

3 Yes, our Epilogue is fiction, set 10,000 years in the future. It's also a vision of a future we are committed to creating.

Appendix A

1 In the equation $O,t=P-S$, the "t" is a mathematical shorthand meaning "over time," which was first popularized by Albert Einstein.

2 C. D. Keeling and T. P. Whorf, "Trends Online: A Compendium of Data on Global Change," Oakridge National Laboratory, 1977, <http://cdiac.esd.ornl.gov/ftp/np001r7/>.

3 COESA, U.S. Standard Atmosphere 1976, USGPO Washington, D.C.

4 See <http://cdiac.esd.ornl.gov/ftp/ndp001/figure/maunalog.gif>. See also <http://cdiac.esd.ornl.gov/ftp/ndp001/maunaloa.co2>.

Appendix B

1 An online sources of of inexpensive peak flow meters is: <http://www.blairex.com/bullet01-02.htm>.

Glossary of Acronyms

AGU	American Geophysical Union
CFCs	Chlorofluorocarbons
CIs	Carbonaceous chondrite meteorites
CMTX	Colliding Micro-Tori Experiment, a desktop fusion device
CO_2	Carbon dioxide
D-He$_3$	Deuterium–helium-3 fusion
EETA79001	Meteorite found at Elephant Morraine in Antarctica in 1979
IMF	International Monetary Fund
JET	Joint European Torus, a fusion device
JPL	Jet Propulsion Laboratory
MGS	Mars Global Surveyor
NASA	National Aeronautics and Space Administration
NEOs	Near-Earth Objects
NORAD	North American Air Defense
NRL	Naval Research Laboratory
NSSDC	National Space Science Data Center
OID	Oxygen Inventory Depletion
OSHA	Occupational Safety and Health Administration (US)
PBFA	Particle Beam Fusion Apparatus
SDI	Strategic Defense Initiative (Star Wars)
SNCs	Category of meteorites which includes the Shergotty, Nakhla, and Chassigny meteorites
SPIT	Starburst Pixel Interleaving Technique
SPSR	Society for Planetary SETI Research
VL-1, VL-2	Viking Landers

Recommended Reading and Resources

Planetary Science

Allan, D. S. and Delair, J. B. *Cataclysm!* (Bear and Company, 1995).

Arms, M. *Riddle of the Ice: A Scientific Adventure into the Arctic* (Doubleday, 1998).

Barbree, J. and Caidin, M. *A Journey Through Time: Exploring the Universe with the Hubble Space Telescope* (Penguin, 1995).

Caidin, M., Barbree, J., and Wright, S. *Destination Mars: In Art, Myth, and Science* (Penguin, 1997).

Carlotto, M. J. *The Martian Enigmas A Closer Look: The Face, Pyramids, and Other Unusual Objects on Mars* (North Atlantic Books, 1997).

DiGregorio, B. E., Levin, G. V. Dr., and Straat, P. A. Dr. *Mars: The Living Planet* (Frog Ltd., 1997).

Fisher, D. E., and Fisher, M. J. *Strangers in the Night: A Brief History of Life on the Other Worlds* (Counterpoint, 1998).

Grinspoon, D. H. *Venus Revealed: A New Look Below the Clouds of Our Mysterious Twin Planet* (Addison-Wesley, 1997).

Lewis, J. S. *Rain of Iron and Ice: The Very Real Threat of Comet and Asteroid Bombardment* (Addison-Wesley, 1996).

McDaniel, S. V. and Paxson, M. R. eds. *The Case for the Face: Scientists Examine the Evidence for Alien Artefacts on Mars* (Adventures Unlimited Press, 1998).

Sagan, C. *Cosmos* (Ballantine, 1980).

Sagan, C. *Pale Blue Dot* (Random House, 1994).

Environmental and Global Warming

Brown, L. R., Renner, M., and Falvin, C. *Vital Signs 1998: The Environmental Trends that are Shaping Our Future* (Worldwatch Institute, 1998).

Caldicott, H., M.D. *If You Love This Planet: A Plan to Heal the Earth* (W. W. Norton and Co., 1992).

Carson, R. *Silent Spring* (Houghton Mifflin, 1962).

Carson, R. *The Sea Around Us* (Oxford University Press, 1950).

Dowie, M. *Losing Ground: American Environmentalism at the Close of the Twentieth Century* (MIT Press, 1995).

Durnil, G. K. *The Making of a Conservative Environmentalist* (Indiana University Press, 1995).

Ehrlich, P. R., and Ehrlich, A. H. *Betrayal of Science and Reason: How Anti-Environmental Rhetoric Threatens Our Future* (Island Press, 1996).

Eldredge, N. *Life in the Balance: Humanity and the Biodiversity Crisis* (Princeton University Press, 1998).

Firor, J. *The Changing Atmosphere: A Global Challenge* (Yale University Press, 1990).

Gelbspan, R. *The Heat is On: The Climate Crisis, The Cover-Up, The Prescription* (Perseus Books, 1997).

Gore, A. *Earth in the Balance: Ecology and the Human Spirit* (Plume, 1993).

Hartmann, T. *The Last Hours of Ancient Sunlight – Waking Up to Personal and Global Transformation* (Mystical Books, 1998).

Meadows, D. H., Meadows, D. L. and Randers, J. *Beyond the Limits: Confronting Global Collapse, Envisioning a Sustainable Future* (Chelsea Green, 1992).

Simon, P. *Tapped Out: The Coming World Crisis in Water and What We Can Do About It* (Welcome Rain Publishers, 1998).

Weisman, A. *Gaviotas: A Village to Reinvent the World* (Chelsea Green Publishing, 1998).

Health and Well-being

Brown, L. R., Renner, M., and Flavin, C. *Vital Signs 1999: The Environmental Trends That Are Shaping Our Future* (W. W. Norton & Co., 1999).

Browner, M. and Leon, W. *The Consumer's Guide to Effective Environmental Choices* (Three Rivers Press, 1999).

Clorfene-Caston, L. *Breast Cancer – Poisons, Profits and Prevention* (Common Courage Press, 1996).

Kubler-Ross, E., M.D. *On Death and Dying: What the Dying Have to Teach Doctors, Nurses, Clergy, and Their Own Families* (Simon and Schuster, 1969).

Peck, M. S. *The Road Less Traveled: A New Psychology of Love, Traditional Values and Spiritual Growth* (Simon & Schuster, 1998).

History

Lord, W. *A Night to Remember* (Bantam Books, 1997).

Resources

To keep up to date with many of the matters introduced in this book or to communicate with the authors, please contact The Garden Earth Enterprise.

The Garden Earth Enterprise
Website: http://www.gardenearth.com
E-mail: gardenert@aol.com

Bibliography

A partial list of publications and reports for which J. E. Brandenburg is listed as author, listed chronologically, is found below:

Monica Rix Paxson, Stephen Corrick, James Erjavec, J.E. Brandenburg, *"All Sinks Plus, The Implications of the Obvious: Rate of Decreasing Atmospheric O2 Constrains Biological Carbon Sink Models?""* American Geophysical Union, Fall Meeting, December14, 1999.

J.E. Brandenburg, *"On the Effect of Large LEO Satcom Constellations on the Lower Thermosphere."* American Geophysical Union, Spring Meeting, June 2, 1999.

J. Erjavec, J.E. Brandenburg, *"Evidence for a Paleo-Ocean Shoreline, Sedimentary Features and Water Erosion in Cydonia,"* American Geophysical Union, Spring Meeting, June 3, 1999.

H. Moore, J.E. Brandenburg, S.K. Corrick, A. Sirisena, *"Ice Found in Craters in Cydonia,"* American Geophysical Union, Spring Meeting, June 3, 1999.

J.E. Brandenburg, J.F. Kline, V.R. DiPietro (Research Support Instruments) *"Progress on the CMTX (Colliding Micro-Tori eXperiment),"* a poster session presented at the 1998 Division of Plasma Physics Meeting of the American Physical Society, Nov. 17, 1998

J.E. Brandenburg, M.R. Paxson, S.K. Corrick, *"OID (Oxygen Inventory Depletion) as a Global Concern: Magnitude, Causes, and Remedies,"* EOS (supplement), Transactions, AGU, Volume 79, Nov. 10, 1998, pg. F172.

J. E. Brandenburg, V. DiPietro, *"The New Mars Synthesis and the Cydonian Hypothesis: Models Confront New Data,"* May 28, 1998, American Geophysical Union Conference.

J. E. Brandenburg, *"Mars as the Parent Body of the Carbonaceous Chondrites and Implications for Mars Biological and Climactic History."* Invited paper. Proceedings of the SPIE, Vol. 3111, pp. 69-80, July, 1997.

J. E. Brandenburg, *"Mars as the Parent Body of the Carbonaceous Chondrites : A Further Examination,"* Meteoritics Conference, Berlin, Germany, 1996.

J.E. Brandenburg, *"Mars as the parent body of the CI carbonaceous chondrites,"* Geophysical Research Letters, Vol. 23, No. 9. pp. 961–964, May 1, 1996

J.E. Brandenburg , *"Constraints on the Martian Cratering Rate Based on the SNC Meteorites and Implications for Mars Climactic History."* Earth, Moon and Planets, pp. 35-45, 1995

J.E. Brandenburg and V. DiPietro, *"Did the Lyot Impact End the Liquid Water Era on Mars?"* American Geophysical Union Conference, Baltimore, MD. June 1995, pg. 52A–3.

J.E. Brandenburg and Michael Micci, *The Microwave-Electro-Thermal Thruster: A New Technology for Satellite Propulsion and Attitude Control* Proceedings of the 9th Annual AIAA/USU Conference on Small Satellites, Logan, Utah, Sept. 1995

J. E. Brandenburg *Constraints on the Martian Cratering Rate Based on The SNC Meteorites and Implications for Mars Past Climate* Proceedings of the International Conference On Comparative Planetology, Pasadena, California, June 1994

J. E. Brandenburg *A Model Cosmology Based on Gravity-Electromagnetism Unification* Astrophysics and Space Science 227:133-144, 1995

H. W. Smathers, D. M. Horan, J. G. Cardon, E. R. Maleret, M. R. Corson, and J. E. Brandenburg *Ultraviolet Plume Instrument Description and Plume Data Reduction Methodology* Naval Research Laboratory Report, NRL/FR/8121—93—9531, May 1993

H. W. Smathers, D. M. Horan, J. G. Cardon, E. R. Maleret, L. Perez, T. Tran, J. E. Brandenburg and R. R. Strunce, Jr., *UVPI Imaging From the LACE Satellite: The Starbird Rocket Plume* Naval Research Laboraory Report, NRL/FR/8105—93—0546, August, 1993

H. W. Smathers, D. M. Horan, J. G. Cardon, E. R. Maleret, M. Singh, T. Sorenson, P. M. Laufer, M. R. Corson, J. E. Brandenburg, J.A. McKay, and R.R. Strunce, Jr., *Ultraviolet Plume Instrument Imaging from the LACE Satellite: The Strypi Rocket Plume* Naval Research Laboratory Report, NRL/FR.8121-93-9526, September 1993

H. W. Smathers, D. M. Horan, J. G. Cardon, E. R. Maleret, L. Perez, T. Tran and J. E. Brandenburg *UVPI Imaging from the LACE Satellite: The Low Cost Launch Vehicle (LCLV) Rocket Plume* Naval Research Laboratory Report,

NRL/FR/8121-93-9545, September 1993

H. W. Smathers, D. M. Horan, J. G. Cardon, E. R. Maleret, L. Perez, T. Tran and J. E. Brandenburg *UVPI Imaging from the LACE Satellite: The Nikha Rocket Plume* Naval Research Laborary Report, NRL/FR/8121-93-9537, July 1993

J. E. Brandenburg *Unification of Gravity and Electromagnetism in the Plasma Universe* IEEE Transactions on Plasma Science, Vol. 20, 6, p 944, December 1992

J. E. Brandenburg *The Cydonian Hypothesis* Journal of Scientific Exploration, Vol. 5, 1, pp 1-25, 1991

J.E., Brandenburg, V. DiPietro and G. Molenaar, "*The Cydonian Hypothesis.*" Journal of Scientific Exploration, Vol. 5, No. 1. 1991.

J. E. Brandenburg *A Theory of The Relaxation of Tokamak Discharges* Journal of Plasma Physics, June 1991

J. E. Brandenburg, Robert Seeley, W. Michael Bollen, and Khan Nguyen *The DIMEX Experiment* Mission Research Corporation Report MRC/WDC-R-R230, July 1990

J. E. Brandenburg *An Analytic Model of a Diamagnetic Migma* APS Bulletin "31,"1557, October 1986

J. E. Brandenburg and Richard F. Post *Analytical Field Reversed Equilibria Derived from Self-Consistent Particle Orbits, Part II* Nuclear Fusion "26," 1073, 1986

J. E. Brandenburg and Larry Ludeking *The LMGE (Leaky Magnetized Gap Equation) Model of the Magnetron* Mission Research Corporation Technical Report, MRC/WDC-R-118, August 1986

Bruce Goplen, J. E. Brandenburg and Richard Worl *Canonical SGEMP Simulation Problems* Mission Research Technical Report, MRC/WDC-R-109

J.E. Brandenburg, *The PaleoOcean of Mars*, MECA symposium on Mars: Evolution of its Climate Atmosphere LPI Tech Rep. 87-01, 1986, pp. 96–98 Lunar and Planetary Inst. Houston TX.

J. E. Brandenburg and John Ambrosiano *Physical Properties and Mathematical Models of Foils in High Current Switches* Mission Research Corporation Technical Report, MRC/WDC-R-106, March 1986

J. E. Brandenburg and Robert E. Terry, *The Eroding Foil Switch (EFS) Model* Defense Nuclear Agency Technical Report, DNA-TR-86-083, January 1986

Bruce Goplen, J. E. Brandenburg, and Tim Fitzpatrick *Transmission Line Matching in MAGIC* Mission Research Corporation Report, MRC/WDC-R-102, September 1985

J. E. Brandenburg *A Simple Model of Conductivity Channel Motion During Electron Beam Hose Instabilities* Lawrence Livermore Laboratory Report UCRL-87096, submitted to Applied Physics Letters, November 1985, with E. P. Lee

J. E. Brandenburg *A Model of Hose Instabilities in Rotating Electron Beams* submitted to Phys. Fluids, August 1985, with E. P. Lee

J. E. Brandenburg *AXFOIL: A Simple Finite-Element Computer Code to Model Moving Axial Foils* Berkeley Associates Technical Report, BRA-85-310R with John Ambrosiano, July 1985

J. E. Brandenburg *A Model of Hose Instabilities in Intense Electron Beams Including the Effects of Return Current and Conductivity Channel Dynamics* APS Bulletin "29," 1291, October 1984

J. E. Brandenburg *A Physical Derivation of Resistive Hose Instabilities in Electron Beams with Return Current* Sandia Report, SAND-84-1026, August 1984

M. A. Sweeney, R. A. Gerber, D. J. Johnson, J. M. Hoffman, P. A. Miller, J. P. Quintez, J. E. Brandenburg, and S. A. Slutz *Analysis of Mechanisms for Anode Plasma Formation in Ion Diodes* Proc. BEAMS '83/5th Int. Conf. on High-Power Particle Beams, San Francisco, CA., September 1983

G. R. Allen, H. P. Davis, and J. E. Brandenburg *Thomson Scattering Measurements of Electron Temperature and Density in A Plasma Channel Created by a Relativistic Electron Beam* Proc BEAMS '83/5th Int. Conf. on High-Power Particle Beams, San Francisco, CA., September 1983

J. E. Brandenburg *A Simple Model of Hose Instabilities in Rotating Electron Beams* presented at 1983 IEEE Int. Conf. on Plasma Science, San Diego, CA., May 23-25, 1983

J. E. Brandenburg *A Theory of the Relaxation of Tokamak Discharges* presented at 1983 Sherwood Theory Meeting Annual Controlled Fusion Theory Conference, Arlington, VA., March 21-23, 1983

J. E. Brandenburg *A Theory of The Relaxation of Finite Beta Toroidal Plasmas* Lawrence Livermore Laboratory Report UCRL-87096, March 1983

303

M. A. Sweeney, J. P. Quintenz, J. A. Halbleib, J. E. Brandenburg and G. W. McClure *A Model for Surface Flashover Sources in Grooved Anodes* APS Bulletin "27," 124, October 1982

J. E. Brandenburg *A Theory of The Relaxation of A Field Reversed Theta Pinch* Lawrence Livermore Laboratory Report, UCRL-86431, October 1982, Submitted to Phys. Fluids

J. E. Brandenburg *Radlac-II Beam Extraction* with J. W. Poukey, and J. R. Freeman, DARPA/Services Propagation Review, SRI International, LLNL, June 21-24, 1982

R. F. Post and J. E. Brandenburg *Analytical Field Reverse Equilibria Derived from Self-Consistent Particle Orbits, Part I* Nuclear Fusion "21," 1633, 1981

Index